AF539418

MOLECULAR DEVELOPMENTAL BIOLOGY

ENCYCLOPAEDIA OF DEVELOPMENTAL BIOLOGY-IV

MOLECULAR DEVELOPMENTAL BIOLOGY

By

Manju Yadav

Lecturer

Department of Zoology

M.M.H. College

Ghaziabad (U.P.)

(India)

DISCOVERY PUBLISHING HOUSE PVT. LTD.

NEW DELHI-110 002

First Published-2008

ISBN 978-81-8356-300-0

Published by

DISCOVERY PUBLISHING HOUSE PVT. LTD.

4831/24, Ansari Road, Prahlad Street,
Darya Ganj, New Delhi-110002 (India)
Phone: 23279245 • Fax: 91-11-23253475
E-mail: dphbooks@rediffmail.com
dphtemp@indiatimes.com

Printed at:

Sachin Printers, Delhi

Preface

The present title **Molecular Developmental Biology** is an important link between almost all biological sciences, and is assuming more and more importance in fields such as medicine and agriculture. Molecular biology has began fulfilling its long anticipated role of linking genetics and embryology and many questions of embryology that had lain dormant for decades have been taken up with new molecular tools and strategies. While our knowledge of the molecular aspects of development has so drastically increased during the time, our applications of developmental phenomenon in evolution and ecology has also expanded. It is becoming impossible to study any area of biology without some background in developmental biology.

This title integrates the descriptive, experimental and biochemical approaches into a conceptual framework for the analysis of development. All important points are illustrated diagrammatically. It is well balanced, completely accessible presentation of principles and application of embryology. In order to create a book which is written within the economy as well as the physical group of the student, only the main essentials of each topic have been presented. It does this with the aid of careful selected examples—some recent and other classic of the field and with numerous illustrations. The aim is to enthuse the reader with this active and exciting area of research and to lay a solid foundation on which further study of its various facts may be based.

To make the work more comprehensive and informative, the author has consulted many authoritative books, research journals, abstracts, monographs etc. He is grateful to all those great scholars whose work are cited or substantially reproduced.

There can be no claim to originality except in the manner of treatment and much of the information has been obtained from the books and scientific journals available in the different libraries.

The author expresses his thanks to his friends and colleagues whose continue inspirations have initiated him to bring out this book.

The author expresses his gratitude to Mr. Wasan and staff of M/s Discovery Publishing House for their whole hearted co-operation in the publication of this book.

Author

CONTENTS

1

Introduction

Embryology, the science of development, has like all other sciences its own embryology. How and when it was conceived remains unknown. Those very early days have been well described by the founder of chemical embryology, Joseph Needham, a gifted historian as well as a biochemist and embryologist. After man discovered artificial incubation of hen's eggs, which was used already in the oldest civilizations for very practical purposes, he must have had sufficient curiosity to break the shell immediately and look at the developing chick embryo. Without knowing it, he became a "descriptive embryologist" and studied morphogenesis (the appearance of new forms and functions). When he made the same kind of observations on a duck embryo, he became a "comparative embryologist". The first known descriptive and comparative embryologist was Aristoteles, who, more than 20 centuries ago, described the development of chick embryos and compared this development with that of embryos from fishes, reptiles, etc. Descriptive and comparative embryology remain at the root of modern embryology. These two branches of embryology, which grew so successfully in the eighteenth and nineteenth centuries, may now seem old; but they are by no means dead, especially in view of the continuous progress made in the optical means of observation. Thanks to the advent of electron microscopy, numerous and valuable papers have been published describing the ultrafine structure of normal sea urchin, frog or chicken embryos. This is the modern form of descriptive and comparative embryology.

However, after man had a good look at the developmental steps that lead to the formation of the adult, he became more curious than ever. How does this organism develop? In the eighteenth century, this question aroused great interest among natural philosophers: some held the view that the adult is already present, in miniature, in the egg *(preformation* theory). A minority, which was admonished by the religious authorities of those days, believed that embryogenesis is an *epigenetic* process; the embryo would form in a progressive way, as the result of complex dynamic changes. This battle was won by the epigeneticists and is now over; however, we shall, in this book, briefly discuss what preformation might mean in molecular terms.

After discussing at length preformation and epigenesis, embryologists were no longer satisfied with the mere observation of developing eggs and began experimenting on eggs and embryos: localized regions of eggs were killed by pricking, fertilized eggs were centrifuged, etc. *Experimental embryology* was horn at the beginning of the present century and culminated in the discovery of Spemann's "organizer": his experiments, demonstrated that the nervous system of a frog does not result from the self-differentiation of a region of the embryo, but from interactions between this region and the underlying cells, i.e. these cells *induce* the differentiation of the undifferentiated ectodermal cells into nerve cells. In addition to induction, other important theoretical concepts, such as potentialities, determination, regulation, germinal localizations, morphogenetic fields and gradients progressively emerged.

Since *chemical embryology is* nothing more than the biochemical analysis of embryonic development, it naturally followed the same trends as embryology itself and its development has been necessarily linked to that of embryology. Chemical embryology began as a descriptive and comparative science and was later deeply influenced by the progress of experimental embryology. It is now being exposed to the impact of bacterial genetics and molecular biology. The use of new methods and concepts originating from molecular biology has transformed "chemical" into "molecular" embryology.

The first important landmark in the history of chemical

embryology was Needham's *Chemical Embryology* (1931). Not only did Needham coin the name of the new science, but he assembled in three large volumes everything that was known at the time about the chemical composition of embryos at all stages of their development. This was the *opus magnum* of a great modern humanist and a bible for the small group of chemical embryologists who were just beginning research at that time. In hundreds of tables, data concerning the protein, glycogen, lipid and water content of all sorts of embryos were compiled from multitudinous papers. This compilation, which was so valuable at the time of its publication, has lost most of its usefulness now, since both problems and methods have undergone considerable changes.

Recent progress made in biochemistry and in physical chemistry had, of course, much influence on the orientation of the more modern work in the 1990s. While reinvestigating Warburg's former observations on the increase of oxygen consumption during fertilization in sea urchin eggs, Runnstrom introduced for the first time into these studies some very important biochemical factors (cytochrome oxidase, cytochromes, dehydrogenases and adenosine triphosphate). It was in the 1980s, too, that Needham, Chambers, Rapkine, Ephrussi, Wurmser and Reiss tried to measure the pH and the redox potential of these eggs. In view of the complexity of the ultrastructure of all cells, and eggs in particular, we can no longer consider a cell as a water solution surrounded by a lipid membrane; it therefore seems foolish to try to determine the internal pH or redox potential of such a complex object as a whole, intact egg. Yet, as we shall see, changes in internal pH seem to play a very important role in the control of protein synthesis at fertilization. It was found by Chambers that the nucleus of an oocyte neither reduces nor oxidizes injected redox dyes; more recent studies have shown that the cell nucleus is indeed remarkably poor in oxidizing and reducing enzymes.

In 1999, a revolution occurred, namely the discovery by Bautzmann et al. that a killed organizer is still capable of inducing a nervous system after grafting it into an early frog or newt embryo. Biochemists discovered the existence of experimental embryology and the whole course of chemical embryology changed. This shift in interests is obvious when one compares Needham's *Chemical Embryology* (1999) with his *Biochemistry and Morphogenesis* (2001) or the senior author's *Embryologie chimique* (2002).

As soon as Holtfreter (1998) and Wehmeier (1999) found that the "inducing substance (or "evocating substance" as the Cambridge group, following Waddington's suggestion, preferred to call it) was of very widespread distribution in Nature, the quest for its identification began. Since an alcohol-treated fragment of horse or ox liver, as well as a killed organizer, "induce" the formation of neural structures in the ectoblast, why not take the whole liver and fractionate it, following the biochemical methods which led to the isolation and purification of vitamins and hormones? This was done in several laboratories and the field was almost as "hot" as that of molecular genetics a few years ago. It soon appeared that every leading laboratory had its own "active substance", differing from that found in other laboratories.

But, suddenly, the setback came and the excitement ceased. In Cambridge, we (Waddington, Needham and Brachet) studied the respiratory metabolism of the organizer as compared with other parts of the gastrula. Since methylene blue was known to increase the respiratory rate of many cells, it was decided to implant pieces of agar containing methylene blue into gastrulae. Neural inductions were obtained, using an "inductor" of *nonbiological* origin. Still worse, neutral red, which has no effect on oxygen consumption, was also active. All the biological substances tested were more or less active; but it was impossible to decide which was the right one, since the ectoblast (like the unfertilized egg treated with parthenogenetic agents) could react in the same way (neural induction) to a large variety of chemical substances. This led Waddington et al. (1996) to suggest that the neuralizing factor was already present in the ectoblast, but in a bound form. Any agent which could "unmask" the inducing substance would induce the transformation of ectoderm into neural tissue. The "true" active substance would be liberated by the dying cells, and the hope of isolating and identifying it became remote. The final blow came when Holtfreter (1997) demonstrated that a short acid or alkaline shock, which is not sufficient to kill the cells, is sufficient to obtain a high percentage of nervous structures in ectoblast explants.

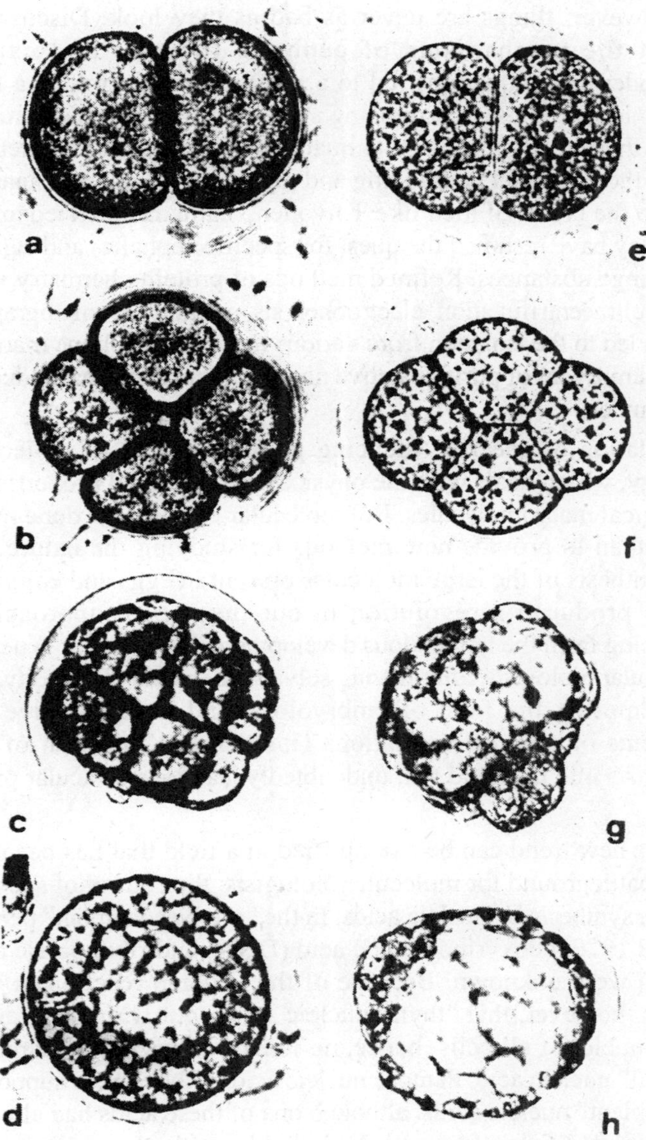

Figure 1.1 : a–h. Early development of sea urchin Paracentrotus lividus (a–d) and mouse (e–h) eggs. a, e 2-cell stage; b, f 4-cell stage; c, g 32-cell stage; d sea urchin blastula. h full-grown mouse blastocyst.

However, things are never as bad as they look. Discussions about the mechanisms of cephalic (neural) and caudal (mesodermal) inductions led to a resumption of work in a field which looked most unpromising since Holtfreter's observations. Were these differences due to quantitative or qualitative factors? Were there distinct neuralizing and mesodermalizing substances? It is to the credit of men like Toivonen, Yamada and Tiedemann that they have resumed the quest for specific cephalic- and caudal-inducing substances. Refined methods of protein chemistry were used (ultracentrifugation, electrophoresis, column chromatography) which led to the isolation from various tissues (liver, bone marrow, chick embryos) of purified active neural and mesodermal-inducing substances.

Isolation of specific, inducing proteins is part of molecular biology, which deals with the physical and chemical properties of biological macromolecules. But molecular biology has done much more than to provide new methods for studying the nature and the synthesis of the large molecules present in eggs and embryos. It has produced a revolution in our theoretical approaches, stemming from the tremendous development of molecular genetics. Molecular biologists, after having solved the riddles of heredity, are now entering the field of embryology and hope to solve the problems of cell differentiation. This massive injection of top biologists into this field will undoubtedly lead to spectacular progress.

This new trend can be exemplified in a field that has been the main battleground for molecular biologists: the control of specific protein synthesis by nucleic acids. In the "pre-Needhamian" period, around 1970, deoxyribonucleic acid (DNA) and ribonucleic acid (RNA) were unknown. Because of the Feulgen reaction, it was known, however, that "thymonucleic acid" (our DNA) is present in the nuclei of all cells, hence, no longer deserving its name of "animal" nucleic acid. Pentose nucleic acids (RNA) were supposed to be "plant" nucleic acids, although one of these acids had already been isolated from pancreas. No reliable method for estimating the two nucleic acids was available, since it was still believed that they could not co-exist in the same cell, one being specific for animal cells, the other for plant cells.

In the case of oocytes, the very presence of thymonucleic acid in the chromosomes was still hotly disputed; discussions between the holders of the "migration" theory and those of the "net synthesis" theory were intense. According to the former, the oocyte nucleus (germinal vesicle) contained a reserve of nucleic acid which moved into the cytoplasm at maturation and migrated back into the nuclei during cleavage. Their opponents, of course, denied the existence of such a reserve and claimed that thymonucleic acid was formed de novo in the nuclei after fertilization. The fact that the very unreliable biochemical methods then available for nucleic acid determination indicated that there was no change during development confused the situation even more. This constancy of the nucleic acid content during development was, of course, the main argument in favor of the migration theory; but the fact that the Feulgen reaction was negative in the oocyte and strongly positive in the nuclei of blastulae and gastrulae apparently demonstrated the correctness of the net synthesis hypothesis.

As usual, it was the development of a new and more specific technique (the diphenylamine method for DNA estimation of Dische) and of new hypotheses that clarified the situation. In 1983, the senior author demonstrated that unfertilized sea urchin eggs contain only traces of DNA as compared to the high DNA content of gastrulae. The reality of a net synthesis of DNA during development thus became an established fact. He also proved at the same time that *eggs* contain large amounts of RNA; therefore, the latter is not a plant nucleic acid at all, but is present in all cells.

A new question arose: where is the intracellular localization of RNA? In 1980, Caspersson's studies with the UV microscope and the senior author's utilization of a simple staining procedure (combined with the digestion of the tissue sections with ribonuclease in order to ensure specificity) brought the desired answer: RNA is an ubiquitous constituent of all cells. It is mainly localized in the nucleolus and the cytoplasm; and there is a close correlation between the RNA content of a given cell and its ability to synthesize proteins.

In vertebrate eggs (those of the amphibians, in particular), RNA is distributed along a primary, polarity (animal-vegetal) gradient already present in the oocyte and the unfertilized egg. At gastrulation, a new gradient, which is more dynamic and results from synthesis of fresh RNA, spreads from the dorsal to the ventral side. As a result of mutual interaction, dorsoventral and cephalocaudal RNA gradients become apparent in the late gastrula and the early neurula stages. These gradients are parallel or identical to the morphogenetic gradients of the experimental embryologists. Disorganization of the gradients by chemical or physical means and abnormalities in development go hand in hand.

The significance of the RNA gradients in amphibian eggs is becoming somewhat clearer. Autoradiography has now shown that RNA synthesis, which leads to the formation of the dorsoventral gradient at the gastrula stage, begins in the nucleus. On the other hand, electron microscopy has shown that the primary (polarity) gradient is essentially a ribosomal gradient. Combining these two findings we can arrive at the following explanation. The polarity gradient of inert ribosomes becomes activated toward gastrulation by messenger RNAs formed in the nuclei along the dorsoventral gradient. The result would be that protein synthesis will be more active in the anterior (cephalic) part of the embryo than in its posterior (caudal) part; protein synthesis would also decrease progressively from dorsal to ventral. Experiments with specific inhibitors of RNA synthesis (actinomycin D) and of protein synthesis (puromycin) confirm these deductions.

This is certainly progress; but to understand how production of specific messenger RNAs, controlled by localized gene activation leading to the synthesis of specific proteins, ultimately leads to the differentiation of nervous system, chorda or muscle cells is a task for the immediate future; it is a task for *molecular embryology*.

2

Specific Proteins in Living Cells

The control of specific protein synthesis is at the core of molecular biology; this problem can be studied in vitro in cell extracts containing enzymes which catalyze specific steps of this complex process. However, since we are dealing in this book with eggs and sperms, mention should first by made of *cell organization;* we shall then discuss specific protein synthesis in living cells. It should be emphasized at once that there is no such thing as a "typical" cell: as we shall see, both eggs and spermatozoa are highly specialized cells. During development, cells, which looked similar, undergo cell differentiation, a necessary step in the diversification of our tissues and organs (brain, muscles, skeleton, etc.). However, since all cells are built on the same morphological pattern.

All cells — except bacteria and mammalian red blood cells - possess *a nucleus* surrounded by a *cytoplasm;* they are limited by a *cell* (or *plasma) membrane* which regulates the exchanges of materials between two adjacent cells and with the outside medium. This membrane is composed of a double layer of lipids, in which "integral" proteins are inserted. Some of these proteins are enzymes which play an essential role in maintaining the cell's chemical composition. For instance, the plasma membrane contains a "sodium pump" which expulses excess sodium ions from the cell and allows a simultaneous intake of potassium ions. Outside the plasma membrane is the *extracellular matrix,* made of sugar-containing proteins (glycoproteins) and sulfated polysaccharides; its main function is to allow the cells to adhere to their neighbors or to an artificial substratum (glass, plastic, etc.). Just under the cell membrane is the *cell cortex:* its main constituent is a bundle of circular *microfilaments* made of *actin.* This protein can undergo linear polymerization or depolymerization by binding to other proteins; calcium ions are often required for binding between actin and the various actin-binding proteins, in particular to the contractile protein, myosin. The final result of this actin polymerization-depolymerization process is that cells can change their shape and undergo locomotion. In a tissue, cells are held together in close contact by various kinds of junctions, thus allowing the passage of small molecules and electrical current from cell to cell.

In the *cytoplasm* we find, as one of the main constituents, *a cytoskeleton* formed by a complex meshwork of fibrous molecules. There are three main cytoskeletal components: *microtubules* resulting from the linear polymerization of a protein called *tubulin;* actin *microfilaments* similar to those which are accumulated in the cell cortex and which may display, if actin binds to myosin, contractility. *Actomyosin* is the major contractile protein of our muscles, but it is also present in almost all cells, including eggs. Finally, *intermediary filaments,* smaller in diameter than the microtubules and larger than the microfilaments, have a more diversified composition: they are made of fibrous proteins which differ in a tissue-specific manner. There are continuous interactions between the three kinds of cytoskeletal elements; the cytoskeleton is thus a dynamic, always changing, intracellular structure and lacks the rigidity of our own skeleton.

In the meshes of the cytoskeletal network, many organelles can be found. Of particular importance are the *mitochondria* and the *endoplasmic reticulum.* The former are surrounded by a double membrane and contain all the enzymes necessary for the oxidation of oxidizable substrates; their oxidation produces the energy required for the synthesis of macromolecules and for cell locomotion. However, cells cannot directly use the energy provided by cellular oxidations: this energy must be stored, in the form of energy-rich phosphate bonds, in a nucleotide called *ATP*. Its structure can be written as follows: adenine ~ ribosephosphate ~ phosphate ~ phosphate, where the sign ~ represents an energy-rich bond. When the cell requires energy, the terminal phosphate bond is broken down by enzymes called *A TPases*. The *endoplasmic reticulum* is made of membranes (similar to the plasma membrane) associated to the *ribosomes,* which will be discussed later in this chapter; their function is protein synthesis.

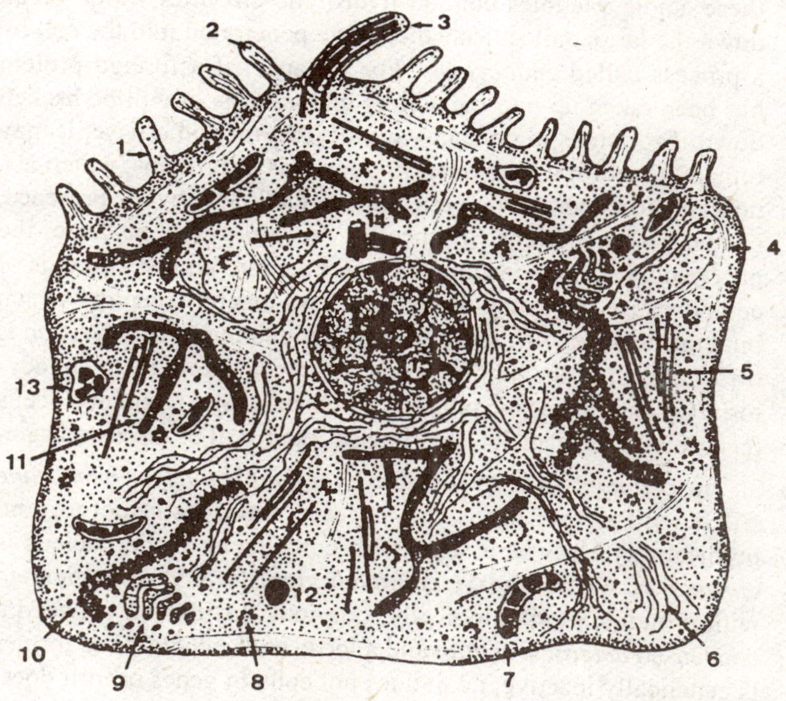

Figure 2.1 : A "typical" cell. I plasma membrane; 2 microvillus; 3 cilium: 4 microfilaments; 5 microtubules; 6 intermediate filaments; 7 mitochondrion; 8 polyribosome; 9 dictyosome (Golgi apparatus); 10 rough endoplasmic reticulum; 11 smooth endoplasmic reticulum; 12 primary lysosome; 13 secondary lysosome; 14 centrioles; 15 nucleus.

The newly synthesized proteins can be either secreted out of the cell or retained inside the cell. Secretory proteins are produced by the endoplasmic reticulum, cellular proteins by free ribosomes (not attached to membranes). The role of the *Golgi bodies,* also found in all cells, is more complex; they contain enzymes which add sugar residues to the proteins synthesized by the endoplasmic reticulum. These sugars constitute "tags" for the various proteins which will be directed to their proper destination in the cell (as luggage is sent to its destination after being tagged in a railway station or an airport). Finally, we should mention the *lysosomes:* these acidic vacuoles contain hydrolytic enzymes which break down the large molecules which have penetrated into the cell by a process called *endocytosis.* For instance, if a foreign protein has been taken up into the cell by endocytosis, it will be broken down if it comes in contact with a lysosome; otherwise, it may remain intact for a long time. Endocytosis of proteins, which is a normal and important process, may be facilitated by the presence, on the cell membrane, of specific *receptors.* For instance, the pancreatic hormone insulin can penetrate into almost all kinds of cells because their membrane possesses insulin receptors which bind the hormone in a specific way; stimulation of cell division is often a result of the binding to receptors and subsequent endocytosis. Other cytoplasmic organelles, which are not found in all cells (centrioles. cilia, flagella, chloroplasts, etc.), will be discussed later.

The cell *nucleus* is surrounded by *a double nuclear membrane* filled with *pores;* more will be said about its structure and permeability elsewhere. The main constituent of the nucleus is *chromatin,* where *deoxyribonucleic acid* (DNA) is associated with proteins; morphologically, one may distinguish between a condensed *heterochromatin* and a loose *euchromatin.* The former is genetically inactive, i.e. it does not contain genes or, if it does, they are in an inactive form; its main role seems to be to keep the chromatin structure intact. In contrast, euchromatin contains active genes which direct, as we shall see, the synthesis of thousands of different proteins in the cell. In addition, most cells - but not all - have a variable number (in general, two) of *nucleoli.* These round bodies are made, as Caspersson and Brachet showed independently in the early 1940s, of *ribonucleic acid* (RNA) and proteins.

Their function will be examined in Chap. IV. Recent work has further shown that the nucleus possesses, like the cytoplasm, a kind of skeleton. This is the *nuclear matrix,* a network of proteins which remains intact after the nucleic acids have been artificially removed from nuclei; long chromatin fibers are wrapped around this nuclear matrix.

After this very brief description of the cell morphology, it is time to discuss a few aspects of its biochemistry, in particular, how genes localized in chromatin direct the synthesis of proteins on cytoplasmic ribosomes.

It is impossible to do more than to summarize here, in a few pages, the essentials of the astounding discoveries made by molecular biologists during the past 40 years. The field is so large and is still moving so rapidly that only the basic facts required for the understanding of molecular embryology can be presented.

It has been known, for very many years, that *proteins* and *nucleic acids* are constituents of all cells, whether they belong to animals, plants or bacteria. Nucleic acids and proteins are also the main (and often the only) constituents of the viruses. These nucleoprotein particles can, if they are introduced into a living cell, replicate themselves repeatedly. When they do so. they retain all their genetic characteristics. If virus A_x is introduced into a suitable cell, many copies of the same A_x viral particles will be formed. This means that viruses, like cells and organisms, have *genetic continuity,* which is the main characteristic of life. The molecular basis of genetic continuity, as we shall now see, always lies in nucleic acids. For this reason, we cannot imagine any form of life, even very primitive, without nucleic acids and proteins. Their main functions are *replication* (one molecule of nucleic acid produces two identical molecules of the same nucleic acid) and *control of protein synthesis.*

The *proteins* are made of amino acids (R—CH—COOH),
                                                    |
                                                   NH_2

linked together by peptide bonds (—NH—CO—) in order to form *polypeptide chains.* Since there are 20 different amino acids and since their position can take any place in the polypeptide chains, an almost infinite number of proteins could theoretically exist.

Figure 2.2 : Spatial (ternary) structure of the polypeptide chain of a hemoglobin molecule.

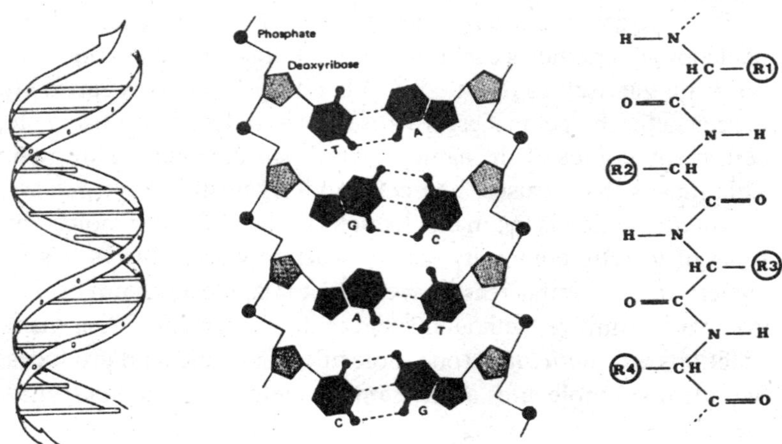

Figure 2.3 : Left: The DNA double helix, according to J. Watson, F. Crick and M. Wilkins. Middle: a segment of the DNA double helix; the bases adenine (A) and thymine (T) on one strand, and guanine (G) and cytosine (C) on the other, are linked by hydrogen bonds. Right: Segment of a polypeptide chain; R1, R2, R3, R4: side chains of the amino acids which form the polypeptide.

The molecular weights of the proteins may vary from a few thousand to several hundred thousand (usually between 10,000 and 500,000) daltons. The largest proteins are often made by the association of a number (usually 2 or 4) of smaller *subunits,* which may be identical or different. The amino-acid sequence of the polypeptide chain defines the *primary structure* of the protein. But the polypeptide chain quickly folds and takes a tridimensional structure, which can be analyzed by X-ray diffraction crystallography. Present methods give us the possibility of establishing not only the amino-acid sequence of the peptide chains, but also the precise spatial structure of such a complex protein as hemoglobin. This molecule, which has a molecular weight of 68,000, contains four polypeptide chains which are identical in pairs (α_2 , β_2). The hemoglobin molecule is abnormal and will not function properly if only one single amino acid of one of the polypeptide chains is replaced by another. Abnormal hemoglobins are far from rare in man, as a result of mutations which give rise to congenital diseases (a gene that has undergone mutation gives rise to an abnormal protein). It should be added that enzymes, which catalyze the biochemical reactions which occur in all cells, are always proteins. Recent research shows that they are often made of nonidentical subunits. One of them has the catalytic activity, while the other plays a regulatory role (it controls, in a positive or negative way, the catalytic activity, acting like the accelerator or the brake of a car).

Enzymes fulfill many functions in the cell: some of them, located in the mitochondria, are involved in energy production; others are essential for the synthesis and the breakdown of the macromolecules (nucleic acids, proteins, glycogen, etc.). Particularly important and interesting are the enzymes which exert a regulatory role in the cell metabolism: for instance, *adenylate cyclase* and *guanylate cyclase,* two membrane-bound enzymes, control the intracellular content of two cyclic nucleotides, cAMP and cGMP. In general, a high cAMP content inhibits cell division and favors cell differentiation; cGMP has the opposite effects. There is currently a great deal of interest in the *protein kinases,* which transfer phosphate from ATP to an acceptor protein; protein phosphorylation modifies the spatial structure and the electrical charge of the protein. Protein kinases, as a rule, require either cAMP, cGMP or calcium ions and phospholipids for activity.

Nucleic acids are, at first sight, much simpler molecules. Instead of 20 amino acids, they vary by the proportion of only four nitrogenous bases (two purines and two pyrimidines). The fundamental unit, in the nucleic acids, is the nucleotide, made of phosphoric acid, a pentose sugar, and a nitrogenous base. There are two main kinds of nucleic acids, which differ by the nature of their sugar moiety: the *deoxyribonucleic acids* (DNA) contain deoxyribose ($C_5H_{10}O4$), while the *ribonucleic acids* (RNA) contain ribose ($C_5H_{10}O_5$). The two types of nucleic acids have three bases in common: the purines adenine (A) and guanine (G), the pyrimidine cytosine (C), while they differ in the fourth pyrimidine base; DNA contains thymine (T), RNA, uracil (U); but thymine is simply a methyluracil. Nucleic acids, like proteins, form long chains. They are, of course, not polypeptide, but *polynucleotide chains*. Figure 2.3 shows a comparison of a polypeptide and a polynucleotide chain.

For many years, noone thought that there could be a link between nucleic acids and proteins. In fact, as late as 1930, DNA was believed to be a small molecule, only present in the nuclei of animal cells, where it would act as a buffer. RNA was supposed to exist only in plant cells, where its possible biological role remained unknown.

A new era, that of molecular biology, started at the time of World War II when three important findings were made.

1. Cytochemical work by Feulgen, by Caspersson and by Brachet showed that DNA and RNA are constant constituents of all kinds of cells. DNA is primarily localized in the cell nucleus and in the chromosomes of the dividing cells. RNA is present in the nucleolus and in the cytoplasm of all cells. Furthermore, a close correlation was observed to exist between the RNA content of a cell and its ability to synthesize proteins. This correlation implied that RNA must play a major role in protein synthesis.

2. Avery discovered the phenomenon of *bacterial transformation.* He found that bacteria (pneumococci), unable to form a protective capsule, could build one if given DNA extracted from bacteria having such a capsule. Since formation or absence of a capsule is a hereditary characteristic, it followed that the genetic material was DNA. In other words, the genes, which are the chromosomal units of heridity, must be made of DNA.
3. The work of Ephrussi, Beadle and Tatum showed that individual genes control the synthesis of particular proteins. This is the "one gene-one enzyme" theory.

Since the gene is made of DNA and since an enzyme is a protein, it follows that the synthesis of an enzyme (that is, the assembly of a specific polypeptide chain) must be controlled by a definite sequence of nucleotides in the corresponding DNA macromolecule. Since, on the other hand, RNA is necessarily involved in protein synthesis, one cannot escape what Crick has called the "central dogma" of molecular biology: *DNA makes RNA, and RNA makes protein.*

How this happens is now well established because of many experiments largely carried out on bacteria and bacteriophages (viruses which infect bacteria). We now know that DNA molecules are made of a giant double helix, where the bases adenine and thymine (A: T) and guanine and cytosine (G: C) are linked together by hydrogen bonding. This structure, as we shall see later, provides an easy explanation for DNA *replication* prior to cell division; it also explains *gene mutation,* which can be due to substitution of one base by another or any kind of defect which will interfere with perfectly accurate base pairing.

The A + T/G + C ratio differs from species to species; it can even vary markedly, in the same chromosome, between different stretches of the DNA molecule. Certain sequences are repeated in tandem many times and constitute *satellite DNAs;* other repeated sequences are interspersed, in many copies throughout the whole genome, between *unique* sequences. It is thought that these unique sequences are those which direct the synthesis of specific proteins and thus correspond to structural genes; the repetitive redundant sequences ("selfish DNA") probably play a regulatory or structural role.

Until recently, it was believed that the huge DNA double helix is remarkably stable except under two circumstances: prior to cell division, it replicates and after an injury (for instance a single-strand break), it is repaired. DNA replication and DNA repair are, in general, carried out by different enzymes. But the recent advent of great technical advances in molecular biology, which cannot be discussed here, has modified our views about the structure and stability of the DNA molecules. Briefly, the methods now available allow us to isolate fragments of the giant DNA molecules which contain individual genes (unique sequences) together with adjacent regions of the DNA double helix. Pure genes can now be obtained in sufficient amounts to analyze their structure, that is to establish the exact sequence of their bases. This work, which is in full swing today, has already led to a number of surprises. For instance, the discovery of *mobile genetic elements"* in yeasts, maize, the insect *Drosophila* and probably many other organisms: these DNA sequences, which display similarities with the cancer viruses and with some of the interspersed repetitive sequences, can jump from one chromosome the another; their function, if any, is not yet clear. However, it is known that insertion of a mobile genetic element into a gene induces its mutation. The very existence of these elements shows that the classical view that the DNA content of a nucleus or a chromosome is absolutely constant can no longer be held. Sequencing DNA molecules has led to two other unexpected discoveries. For many genes, the region upstream the origin of the gene (that is, to the left of the gene) contains *regulatory sequences* which promote the transcription of the gene in the corresponding RNA sequences. However, in genes coding for small RNA molecules, the promoter of transcription is not located in regions flanking the gene, but in the middle of the gene itself. Another unexpected finding has been the discovery of *pseudogenes,* sequences which are very similar to those of a gene, but which cannot direct the synthesis of a protein because they are slightly abnormal. There is growing evidence that at least some of these pseudogenes originate from nuclear RNA molecules that have been copied into their DNA counterparts (reverse transcription); if so, genetic information does not always flow from DNA to RNA as required by the "central dogma" of molecular biology. For some pseudogenes, it is apparently RNA which has made DNA during Evolution.

There is increasing evidence for the view that Evolution results from deeper changes in the DNA molecules than the classical "point" mutations (changes at the level of a single base). Base sequencing shows that gene duplication, insertion or deletion of large DNA regions have been major factors in the production of mutations. Such events still occur today, as shown by the insertion of mobile genetic elements into many *Drosophila* chromosomal loci. In *Trypanosomes,* when the infected host raises antibodies against the surface glycoproteins of the parasite, a new surface coat protein entirely different from the previous one is synthesized. This happens because the parasite is capable of duplicating the gene which directs the synthesis of this new surface protein; this extra copy of the gene disappears when it is no longer required.

We know how the information contained in the "structural" genes, which are also called *cistrons* (that is, those unique sequences which direct the synthesis of a given protein or, more precisely, of a given polypeptide chain; the hemoglobin genes, for instance) is used for the synthesis of the corresponding specific protein. This information is *encoded* in the DNA molecule in the form of sequences of three nucleotides *(triplets or codons).* The information contained in the DNA molecule must first be *transcribed* into an almost identical RNA molecule; this is the so-called *messenger RNA* (mRNA), which is a faithful copy of one strand of the DNA gene. The copy is made in the cell nucleus by an enzyme called RNA polymerase II, which synthesize RNA only in the presence of DNA, which acts as *a template.* The mRNA is then released in the cytoplasm, which is the main site of protein synthesis; it contains the same information as DNA, in its codons. For instance, the AAA sequence (three adenines following each other) corresponds to the amino-acid lysine.

Thanks to the *genetic code,* which has been completely deciphered, it is now possible to deduce the amino-acid sequence of a protein from the base sequence of the corresponding gene. This type of work, as well as other experiments, has disclosed a surprising fact: the majority of the genes are "split", that is the coding sequences are interrupted by noncoding sequences; the first ones are called *exons,* the second *introns.* The gene coding for

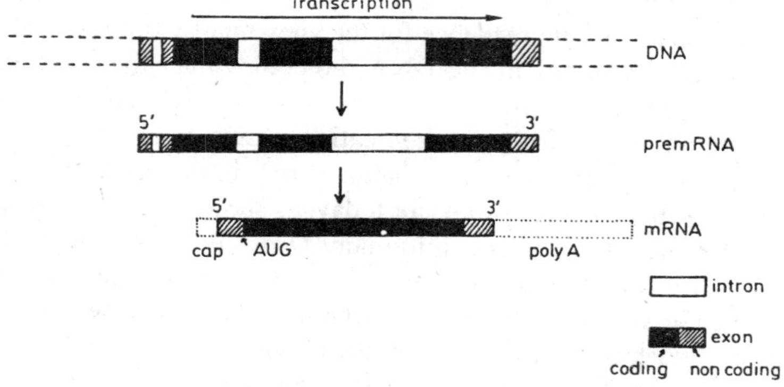

Figure 2.4 : The transcription of an eukaryotic gene. The gene (DNA) is interrupted by noncod ing sequences (introns), which are transcribed and then removed during the maturation of the pre-mRNA into mRNA. In the mature mRNA, a "cap" (7 methyl-guanosine 5' triphosphate, 7 mGppp) has been added at the 5' end and a poly A sequence at the 3' end; noncoding sequences are present at both ends of the molecule. AUG is the initiation codon for translation on the polyribosomes.

one of the collagens is fragmented into as many as 50 introns; in contrast, a few genes have no introns at all. The significance of these mysterious introns is completely unknown; many think that they are just relics of the evolution of ancestral genes. The presence of introns obviously complicates mRNA synthesis: the first product of RNA polymerase activity (the *primary transcript* or *pre-mRNA) is* a RNA copy of the whole gene, introns included. The transformation of this primary transcript into mature mRNA requires the removal of the RNA segments corresponding to the introns by a process called *splicing;* its detailed mechanism is not yet fully understood. It seems certain that the mRNAs cannot move from the nucleus to the cytoplasm unless the *maturation* of the pre-mRNA has been completed in the nucleus. i.e. when all the RNA copies of the introns have been removed.

In a very simplified way, the mechanism of protein synthesis (translation of the genetic code). The specific mRNA molecules bind, in the cytoplasm, to already existing particles, the *ribosomes,* which result from the combination of four kinds of RNAs (the *ribosomal* RNAs or rRNAs) and a large number of ribosomal proteins. Ribosomal RNAs and proteins each form about 50% of the mass of a ribosome; it is customary to describe the ribosomes and the RNAs by their sedimentation constants (S) which can be measured in the ultracentrifuge.

Animal and plant cytoplasmic ribosomes have sedimentation constants of 80S, while bacteria have smaller ribosomes (70S). The cytoplasmic ribosomes, which are by far the most important site of protein synthesis, contain, in animal and plant cells, four different kinds of rRNAs, all rich in GC: 28 S rRNA. which has a molecular weight of about 1.3 million daltons, 18S rRNA with a lower molecular weight of about 600,000 daltons, and the very small 5.8S and 5 S rRNAs. Under appropriate circumstances, the 80S ribosomes dissociate into two *subunits* with S values of 60 and 40, respectively. It is to the smaller 40S subunit that mRNA binds, in such a way, that it links the ribosomes together to form larger aggregates called *polysomes.* There is very good reason to believe that association and dissociation of the subunits is an essential and necessary step in normal protein synthesis.

Why is the larger subunit made? It binds the representatives of the third large class of RNAs, the *transfer-RNAs* (tRNAs). These small RNA molecules (sedimentation constant of *4 S,* molecular weight about *24,000)* are now very well known since the complete nucleotide sequence of many of them has been firmly established. All tRNAs have the shape of a cloverleaf. At the center of the folded molecules is a specific sequence of three nucleotides, the *anticodon,* which binds with the corresponding triplet of the mRNA (the *codon).* If, for instance, the codon in the mRNA is AAA (which, as we have seen, codes for lysine), the anticodon will be UUU (since A and U bind together in the same way in RNA as A and T in DNA).

This particular tRNA will thus be a lysyl-tRNA, capable of

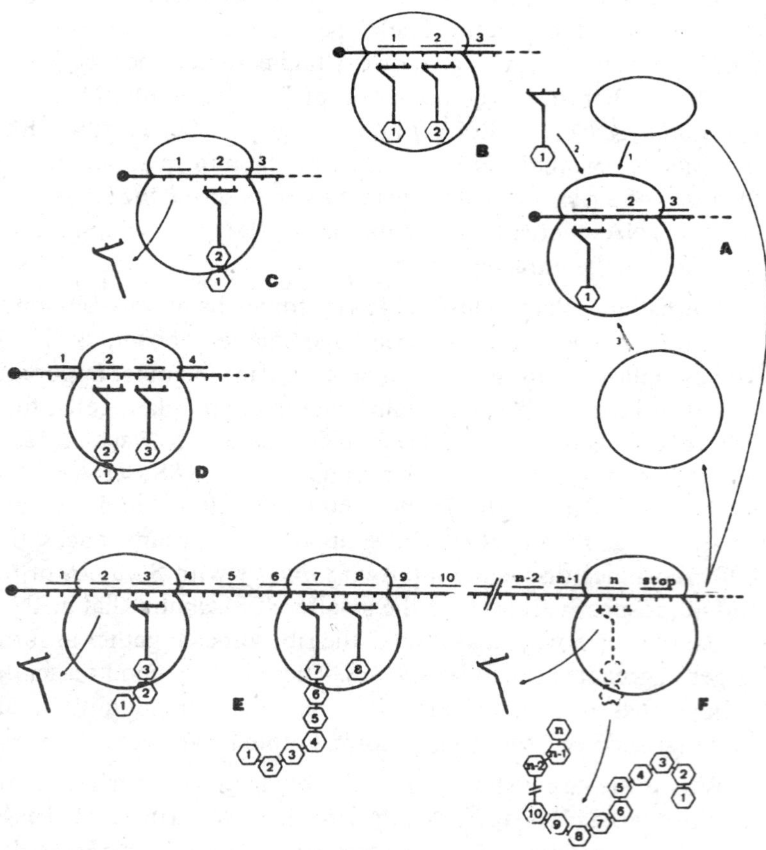

Figure 2.5 A-F. Protein synthesis. A Initiation: Binding of a small ribosomal subunit to a mRNA molecule (1). This is followed by the attachment to the AUG initiation codon in the mRNA of the corresponding aa-tRNA molecule (2) ([1]: formylmethionyl-tRNA). Finally, a large ribosomal subunit binds to the small subunit (3). B Attachment of a second aatRNA molecule [2] to the second codon of the mRNA molecule. C Formation of a peptide bond between aa_1 ***and*** aa_2 ***at the aa-tRNA-binding site. Ejection of the*** $tRNA_1$***. D Movement of the mRNA in order to place the growing polypeptide at the "protein-binding site" and codon 3 at the "aa-tRNA-binding site". Binding of a third aa-tRNA molecule [3] to the third codon of the mRNA molecule. E Elongation of the polypeptide chain. F Ejection of both tRNAn and the synthesized protein.***

accepting the amino-acid lysine at one end of its molecule; this end is made of the same nucleotide sequence (-CCA) in all tRNAs. There are enzymes in the cytoplasm that can remove or add this terminal -CCA sequence, with the result that the tRNA will be uncharged or charged with the corresponding amino acid. Because of the ribosomes and the tRNAs, the genetic message which has been transcribed from DNA to mRNA can be read and *translated* into protein.

It should be added that reality is still much more complex than this scheme. For instance, the genetic code is degenerate; this means that there can be several codons for the same amino acid. Consequently, several "iso-accepting" tRNAs can exist. They have different anticodons, but accept the same amino acid. The proportion and nature of the various iso-accepting tRNAs can change from cell to cell, and even in the same cell when, as the result of hormonal stimulation, for instance, it synthesizes a new protein. Another important factor is that, in order to synthesize a normal, "meaningful" protein, the synthesis should start with the right amino acids and stop when the last amino acid has been introduced in the growing polypeptide chain. Therefore, precise "signals" must exist that control the beginning and the end of the reading of the message. Nonsense codons, that is, nucleotide sequences which do not correspond to any amino acid, provide a means for stopping the growth of the polypeptide chain. The initiation of the latter is very complex indeed. There is not only an initiation codon (AUG) to which a specific tRNA (formyl methionine or methionine tRNA) binds by its anticodon, but at least three initiation factors of protein nature. Such protein factors (which are also needed for the dissociation of the ribosomes into subunits and for the elongation of the growing polypeptide chain) can be considered as specialized enzymes.

We have mentioned that all cells (except bacteria) contain *mitochondria*, which play a most important role in cell respiration and energy production. It is worth pointing out that the mitochondria behave, to a certain extent, like autonomous cell organelles. They contain their own DNA, which has been completely sequenced and is much smaller and entirely different from the main nuclear

DNA. This mitochondrial DNA is replicated when the mitochondria divide and it can be transcribed by a mitochondrial RNA polymerase into all kinds of RNAs. The mitochondria contain their own ribosomes which are even smaller (55S) than those of the bacteria. These ribosomes are capable of independent protein synthesis, probably specialized in the building up of the proteins which form the mitochondrial membranes (structural proteins).

An important tool for the study of protein synthesis (which is very frequently used in molecular embryology) is the use of specific inhibitors of transcription and translation. *Actinomycin* is the best-known *inhibitor* of *transcription.* It binds to DNA and prevents the synthesis of all kinds of RNAs (mRNA, rRNA and tRNA), with a preference for rRNA (which represents 80% or more of the total RNA content of the cell). More recently, it has been shown that RNA synthesis can' also be blocked with *a-amanitin.* This poison inhibits the activity of one of the RNA polymerases, RNA polymerase II, which is localized in the nuclear sap and plays an essential role in the synthesis of the mRNAs. The *inhibitors* of *translation* act at the level of the polysomes, thus in the cytoplasm. The most important are *puromycin,* which provokes the release of incomplete, inactive polypeptide chains and *cycloheximide* which blocks the movement of the ribosomes along the thread of mRNA and thus stops the "reading" of the genetic message.

Mitochondrial protein synthesis is hardly affected by these inhibitors, but is inhibited by the bacterial antibiotics *chloramphenicol* and *tetracycline.* Mitochondrial RNA polymerase, in contrast to the main nuclear enzyme, is unaffected by *a-amanitin,* but it is blocked by an antibacterial and antiviral agent, *rifampicine,* which has no effect on the nuclear RNA polymerases. As one can see, the use of proper inhibitors can allow a discrimination between the synthetic activity of the mitochondria and that of the rest of the cell.

Clearly, mitochondria have much in common with bacteria and it has often been suggested that they once originated from bacteria, and, in the course of evolution, adapted themselves to symbiotic life in animal and plant cells. It should be added that present- day mitochondria are almost entirely dependent on the rest of the cell for their survival and multiplication. They contain so little DNA, as compared to the nucleus, that their genetic information for protein synthesis is restricted to only a small number of proteins. Curiously, the genetic code used by the mitochondria is somewhat different from that of chromatin.

In fact, many of the respiratory enzymes present in the mitochondria are synthesized by the cytoplasmic polysomes under the control of nuclear genes, and move from the cytoplasm into the mitochondria.

More important than the mode of action of specific inhibitors of protein synthesis is the difficult problem of the *control mechanisms* of protein synthesis. We know, from the work of Jacob and Monod, how, in bacteria, gene activity is regulated by cytoplasmic repressors. A protein, the *repressor,* binds with extremely high affinity and specificity, to the DNA of an *operator* gene. When the latter is active (that is, its DNA is not blocked by the repressor), a whole battery of genes (the *operon)* becomes active, leading to the synthesis of the corresponding mRNAs and enzymes.

But it is certain that this scheme, which is valid for bacteria, is insufficient to explain the regulation of genetic activity in more complex cells or organisms. First of all, the nucleus of the bacteria is of the *'prokaryotic"* type; this means that it is extremely simplified. It is made of a circular thread of DNA, the bacterial chromosome, which is not surrounded by any membrane and is probably not associated with proteins. On the other hand, the nucleus of the *eukaryotes* is much more complex. It is surrounded by *a nuclear* membrane or envelope, it contains RNA-rich *nucleoli* and its DNA is associated with many different kinds of proteins to form the *chromatin;* the latter corresponds to the chromosomes of the dividing cells, but it is in a loose, uncondensed form.

There are two classes of chromosomal proteins, the *histones* and the *nonhistone* chromosomal proteins. One finds, in general, five main classes of histones, called H_1, H_2A, H_2B, H_3 and H_4;

but certain cells, like the spermatozoa and the nucleated red blood cells contain other forms of specific histories or histone-like proteins. All the histories are small basic proteins with a high arginine or lysine content. All of them bind strongly to DNA and form particles with it called *nucleosomes*. Histories H_2A, H_2B, H_3 and H, undergo polymerization, forming an octamer; a DNA fiber wraps around this histone *core*. Between two adjacent nucleosomes is *a linker* DNA, which may be associated with histone H,. The length of the DNA molecule which surrounds the core is constant in all species and tissues. On the other hand, the length of the linker DNA may vary in a species- and a tissue-specific manner. The exact number of the nonhistone chromosonal proteins is not known, but it is certainly very high. Very important among them are the *RNA polymerases* I, II and III which are the agents of DNA transcription; they direct the synthesis of the large (28 S and 18 S) rRNAs, the mRNAs and the small RNAs (5S rRNA, tRNAs), respectively. These enzymes cannot work unless they bind to DNA; therefore, in order for transcription to start and to proceed, chromatin should be in a loose, "active" state accessible to the enzymes. We shall discuss later the mechanisms which control transcription; but it may now be mentioned that *active chromatin* differs from inactive chromatin in several respects. It is much more sensitive to deoxyribonuclease (DNase), the enzyme

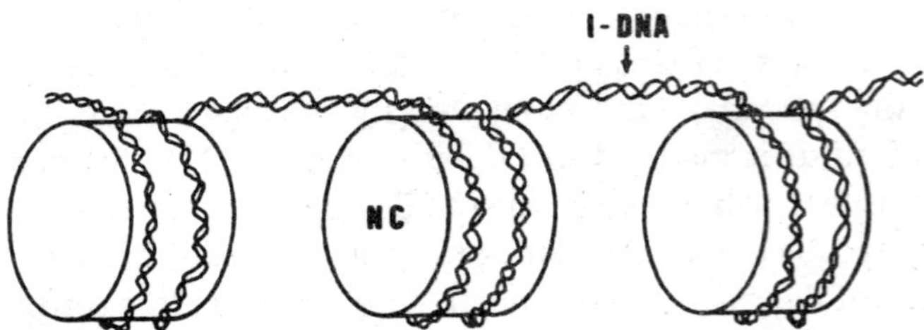

Figure 2.6 : Chromatin organization in nucleosomes. NC nucleosome core (histones); l-DNA linker DNA

which breaks down DNA; it contains different amounts of basic nonhistone proteins (called *HMG proteins)* which, like the histories and the RNA polymerases, bind strongly to DNA. In fact, many nuclear proteins are DNAbinding proteins; among them are enzymes called *topoisomerases* which modify the conformation of the DNA molecules. They contribute to the solution of a major problem for the cell nucleus, that of DNA compaction: DNA is a very long, but very thin molecule. How can it be packed into a nucleus? The first degree of compaction is achieved by the formation of nucleosomes, thus by the binding of DNA to core histories; a higher degree of compaction results from the coiling of the nucleosomes in order to form *supra-nucleosomal structures.*

Finally, it is very likely that the chromatin fibers form loops around the previously mentioned proteinaceous nuclear matrix. As one can see, the very important problem of the control of genetic activity is not yet solved; but we are beginning to see, despite all the complexities of chromatin structure. how to tackle and finally to solve this problem.

Protein synthesis is not only controlled at the transcriptional level. We shall find clear cases, in eggs, of *post-transcriptional* controls, which can occur at many levels: (1) in the nucleus itself, where unwanted mRNA molecules can be destroyed by hydrolytic enzymes (the ribonucleases); (2) at the nucleocytoplasmic border, where a selection between mRNAs which will function as messengers in the cytoplasm and others which will be retained and degraded in the nucleus probably occurs; (3) finally, at the level of the polysomes, in order to modify the translation of a given message. The conformation of the ribosomes, the binding of proteins to the mRNAs, the availability of tRNAs, of activating enzymes for the various amino acids, of initiation factors, etc., could all modify the amount and the nature of the proteins synthesized by a given cell. Since, as discussed in the Foreword, specific protein synthesis and cell differentiation are very closely related events, it is clear that any progress made in the field of gene regulation in eukaryotic cells is of primary importance for molecular embryology.

However, what we really want to know is still more than that: what causes the appearance of typical *morphological* structures, like the cross-striation of the muscle and heart cells, the production of cilia and flagella, which can beat and allow larvae and spermatozoa to swim, etc.? This is the basis of *cell differentiation,* which can easily be recognized under the microscope. It seems, at first sight, that we are dealing with such a formidable problem that improved knowledge of protein synthesis, of control mechanisms, of genetic activity, etc. will be of little use for its solution.

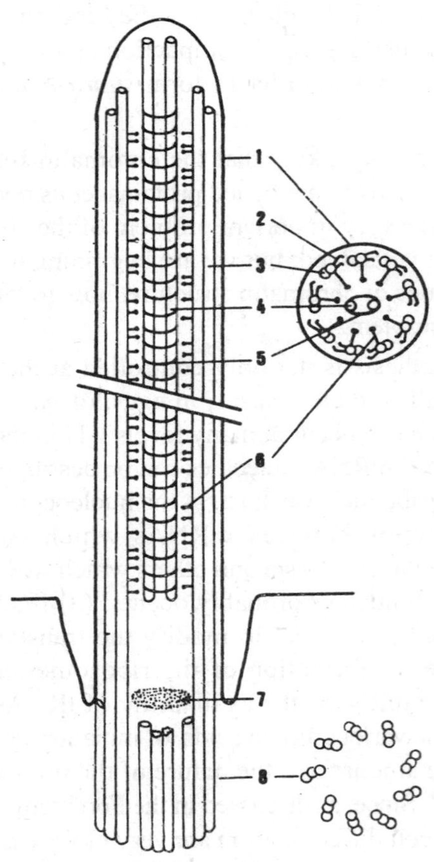

Figure 2.7 : Structure of a cilium (or flagellum) in an eukaryotic cell. **Left:** ***longitudinal section.*** **Right:** ***cross-sections of the cilium and its basal body (kinetosome). 1 ciliary membrane; 2 lateral arms; 3 the nine microtubule doublets; 4 central pair of microtubules; 5 central sheath; 6 radial spokes; 7 ciliary plate; 8 basal body (kinetosome).***

And yet, there are good reasons for hoping that molecular biology and embryology will bring us a great deal closer to this still distant goal. Take the example of the cilia and flagella: they are made of

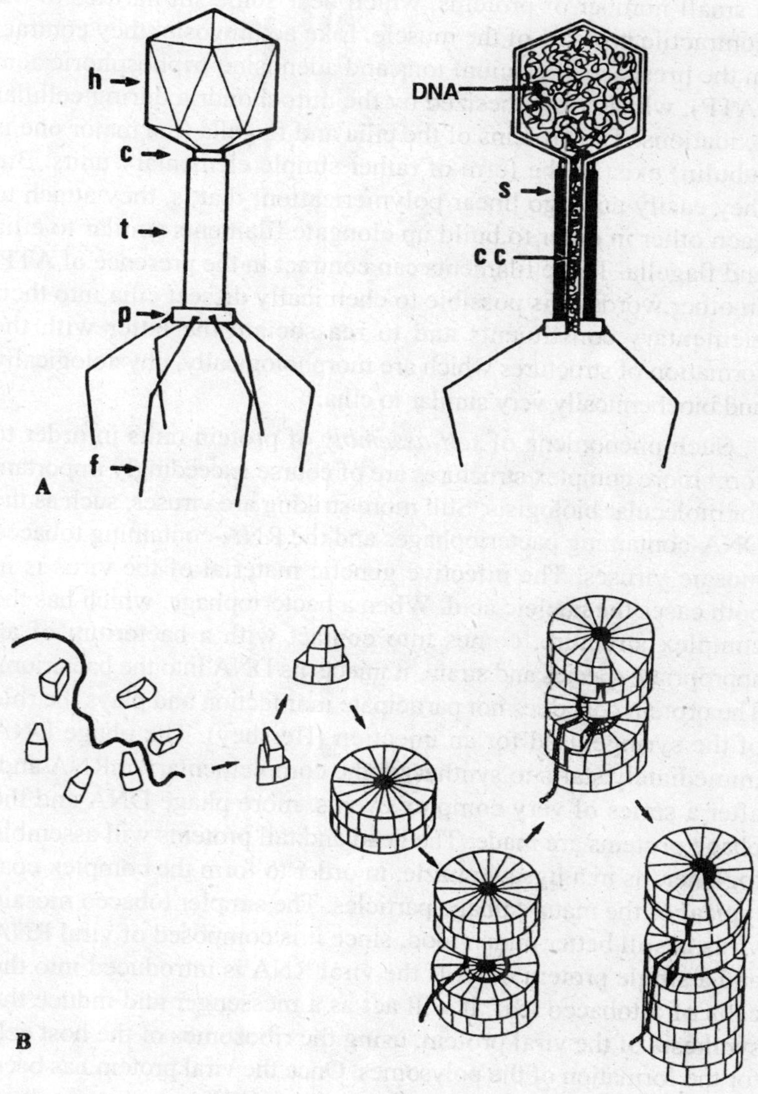

Figure 2.8 A,B. Structure of two viruses. A Bacteriophage (T4-type). Left: schematic view of a virion; h head; c collar; t tail; p tail plate; f tail fibers. Right: ideal section through a virion: cc core with central channel; s sheath. dna DNA molecule coiled inside the head. B Selfassembly of proteins and RNA in tobacco mosaic virus.

a small number of proteins, which bear some similarities to the contractile proteins of the muscle. Like actomyosin, they contract in the presence of calcium ions and adenosine-triphosphoric acid (ATP), which is synthesized by the mitochondria during cellular oxidations. The proteins of the cilia and flagella (the major one is tubulin) exist in the form of rather simple elementary units. But they easily undergo linear polymerization; that is, they attach to each other in order to build up elongate filaments similar to cilia and flagella. These filaments can contract in the presence of ATP. In other words, it is possible to chemically dissect cilia into their elementary constituents and to reassociate the latter with the formation of structures which are morphologically, physiologically and biochemically very similar to cilia.

Such phenomena of *self-assembly* of protein units in order to form more complex structures are of course exceedingly important for molecular biologists. Still more striking are viruses, such as the DNA-containing bacteriophages and the RNA-containing tobacco mosaic viruses. The infective genetic material of the virus is in both cases the nucleic acid. When a bacteriophage, which has the complex structure, comes into contact with a bacterium of an appropriate species and strain, it injects its DNA into the bacterium. The protein coat does not participate in infection and plays the role of the syringe used for an injection (Hershey). The phage DNA immediately starts to synthesize the complementary mRNA and, after a series of very complex events, more phage DNA and the phage proteins are made. The head and tail proteins will assemble together, as in a jigsaw puzzle, in order to form the complex coat typical of the mature phage particles. The simpler tobacco mosaic virus is still better understood, since it is composed of viral RNA and a single protein only. If the viral RNA is introduced into the cells of a tobacco leaf, it will act as a messenger and induce the synthesis of the viral protein, using the ribosomes of the host cell for the formation of the polysomes. Once the viral protein has been formed, complete rods of virus particles will form, as a result of self-assembly of RNA and protein molecules.

Studies on the self-assembly of viral nucleic acid and proteins represent a very useful model for the understanding of the much more complicated morphogenic events which occur during embryonic development. They also inspire the hope that it will be possible, in a not very distant future, to synthesize the RNA and the protein constitutive of the virus and to obtain in vitro, fully infectious tobacco mosaic viral particles.

3

Gene in Determination and Differentiation

During the development of a multicellular organism, numerous functional cell types are produced from a single cell-the fertilized egg, or zygote. The magnitude of this feat and the challenge of explaining development become apparent if we examine the starting material that the zygote utilizes to form a complex organism: a single diploid nucleus, which is embedded in the egg cytoplasm and contains the genes that designate the properties of each and every cell in the body.

How does this single genome produce the myriad cell types that constitute the completed organism? The Roux-Weismann theory proposed that the nucleus contains determinants that are sorted out during mitosis and, by this means, gradually segregated into different cells; the presence of different determinants in different cells was the apparent answer to the question of how cell determination occurs. The cell separation experiments by Driesch were the undoing of Roux and Weismann; clearly, there is no loss of genetic information by early cleavage stage nuclei.

However, the Driesch experiments did not completely quell the beliefs of many developmentalists that modifications in the genetic material itself could occur during development. The task of the nucleus would be much simpler if unnecessary genetic information could be dispensed with or modified in some way so that it becomes nonfunctional. In other words, the nucleus itself would become highly adapted to perform a singular role in development, for example, the production of a blood cell or a bone cell or a skin cell. Such adaptation would commit a nucleus permanently to a restricted developmental role. In the jargon of embryology, the nucleus would lose potency. The zygote nucleus is said to be totipotent—it has complete potential to produce the entire range of cell types that make up the organism. We shall now examine the results of experiments that were designed to test the potency of nuclei of differentiated cells.

GENOME EQUIVALENCE

One way to determine whether the genomes of differentiated cells retain genetic information that is not utilized in those cells is to challenge their nuclei experimentally to reinitiate development and promote differentiation of alternate cell types. The cell separation experiments discussed in Chapter 1 indicate clearly that early cleavage stage nuclei are totipotent. But what happens to nuclei as development continues past early cleavage? Are they irreversibly changed during later development? Cell separation experiments become impractical in later stages, since the volume of cytoplasm is reduced during each cleavage division, and the reduction can affect the ability of nuclei to function properly. Plants are a definite exception to animals in this regard. As we shall see later in this chapter, isolated plant cells-even from adults-may be induced to form an entire plant.

An alternative (and highly ingenious) procedure for analyzing nuclear potential in later cleavage stage newt embryos was developed by Hans Spemann. He partially constricted a *Triturus* egg with a hair loop just before cleavage, confining the nucleus to one half of the egg. A narrow cytoplasmic bridge joined the nucleated and enucleated halves. At the first cleavage the two cells that formed were decidedly unequal. One contained approximately one fourth of the cytoplasm, whereas the other was in the shape of an asymmetrical dumbbell with one fourth of the cytoplasm on the nucleated side and one half of the cytoplasm on the enucleated side. The two lobes of this cell remained in communication via the cytoplasmic bridge, but the nucleus stayed in the smaller lobe, unable to cross the cytoplasmic bridge because it remained too

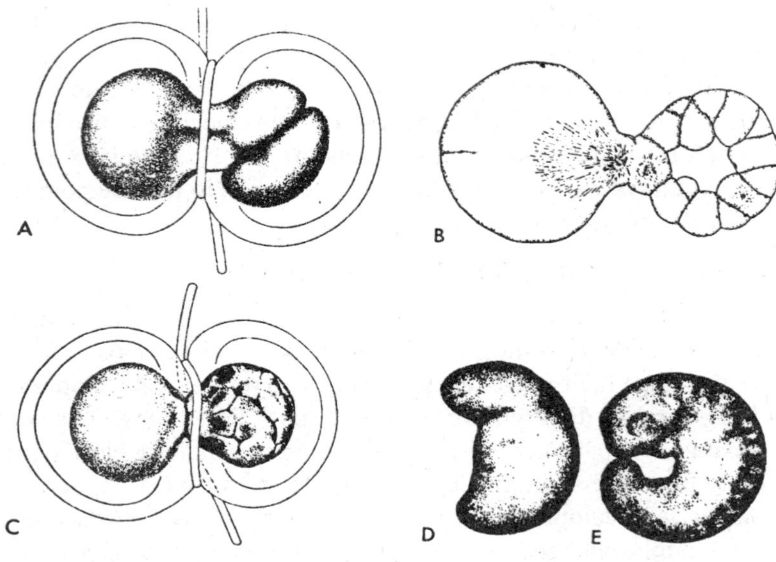

Figure 3.1 : Spemann's delayed nucleation experiment. A, Ligature confines nucleus to right half, which cleaves. B and C, Nucleus adjacent to intracellular bridge enters enucleate half and cleavage furrow appears in bridge. D and **E,** ***Embryos following completion of constriction. The embryo developing from the side with delayed nucleation*** **(D)** ***developed more slowly.*** **(A, C, D,** ***and*** **E** ***from H. Spemann. 1928. Die Entwicklung seitlicher and dorso-ventraler Keimhalften bei verzogerter Kernversorgung. Zeitschrift fiir wissenschatfliche Zoologie, 132: 105-134; B from G. Fankhauser. 1930. Zytologische Untersuchungen an geschnürten Triton-Eiern. 1. Die verzogerte Kernversorgung einer Halfte nach hantelformiger Einschnbrung des Eis. Wilh. Roux' Arch. fur Entwicklungsmechanik der Organismen, 122: 135.***

large to enter. At each subsequent division the small lobe of the dumbbell divided, always forming one small complete cell and one cell that remained continuous with the cytoplasmic bridge. During cleavage, nuclei became progressively reduced in size; by the 16-cell : stage, the nuclei were so small that during the next division, one of the daughter nuclei was able to slip through the cytoplasmic bridge and into the large lobe of the dumbbell. Since a cell membrane forms between daughter nuclei after mitosis, a membrane appeared in the bridge, forming a complete cell in the previously inoculated half.

After renucleation the undivided half of the embryo began to cleave, and Spemann drew the hairloop tight, completely separating the two halves. Both halves continued to develop normally, the renucleated side somewhat more slowly. The nucleus that entered the enucleated half of the embryo was equivalent to a nucleus at the 32-cell stage. Upon separation of the renucleated half of the embryo, the 32-cellstage nucleus was shown to contain the complete set of genetic information for the formation of an entire embryo. Clearly, there was no loss of developmental potential during the cell divisions that led up to the 32-cell stage.

Nuclear Transplantation

In Spemann's constriction experiments a nucleus was transmitted through a bridge of cytoplasm to a region of the embryo that was previously devoid of a nucleus. Several years later, Spemann (1938) suggested a more sophisticated method for placing a nucleus in enucleated cytoplasm and testing its potency:

> Probably the same effect could be attained if one could isolate the nuclei of the morula and introduce one of them into an egg or an egg fragment without an egg nucleus. The first half of this experiment, to provide an isolated nucleus, might be attained by grinding the cells between two slides, whereas for the second, the introduction of an isolated nucleus into the protoplasm of an egg devoid of a nucleus, I see no way for the moment. If it were found, the experiment would have to be extended, so that older nuclei of various cells could be used. This experiment might possibly show that even nuclei of differentiated cells can initiate normal development in the egg cytoplasm.

Spemann was suggesting the possibility of nuclear transplantation from a somatic cell to an enucleated egg but could see no way to perform the experiment. However, 14 years later, Briggs and King (1952) reported successful transplantation of nuclei of the leopard frog, *Rana pipiens*. Their procedure involves two main steps: (1) preparation of the recipient egg and (2) isolation of a donor cell and transfer of its nucleus. The recipient eggs are obtained from a female by induced ovulation. Since normal fertilization has two components-activation of development and formation of a diploid nucleus-successful nuclear transplantation must also mimic these events. Activation of the egg is normally caused by the interaction of the sperm with the egg surface. Activation for nuclear transplantation is achieved by pricking the surface of the egg with a clean glass needle. As a result, the egg nucleus moves to the animal pole in preparation for the formation of the second polar body. The location of the nucleus is indicated by a depression in the egg surface at the animal pole, facilitating its removal with a second glass needle. Donor cells are obtained by dissociating the cells of an embryo in a special solution. Normal cell adhesion requires the presence of calcium and magnesium ions. The dissociation solution contains no Ca^{++} or Mg^{++} and may contain ethylenediamine tetra-acetic acid (EDTA), which binds the Ca^{++} and Mg^{++} located on the surface of the embryo cells, causing them to separate from one another. An individual cell is drawn into a micropipette that has slightly smaller diameter than the cell itself, causing the cell membrane to break. The nucleus, together with its surrounding cytoplasm, is then injected into the activated enucleated egg with the same pipette.

When nuclei from blastulae were transplanted to enucleated eggs, approximately 55% promoted normal cleavage and blastula formation, and 80% of the resulting blastulae continued to develop to the tadpole stage. Of these, 75% reached metamorphosis, at which stage the experiment was terminated. McKinnell (1992) obtained normal postmetamorphic frogs from blastula nuclear transplants, providing additional evidence that there is no reduction of developmental potential in blastula nuclei. In a series of investigations Briggs and King used donor nuclei from endodermal cells of progressively later stages of development in order to examine the possible loss of developmental potential in nuclei. They indeed found that the percentage of transplants showing normal development declined as the age of the donor nuclei increased. Early gastrula nuclei were nearly as effective in promoting normal development as the cleavage nuclei, but nuclei from late gastrula

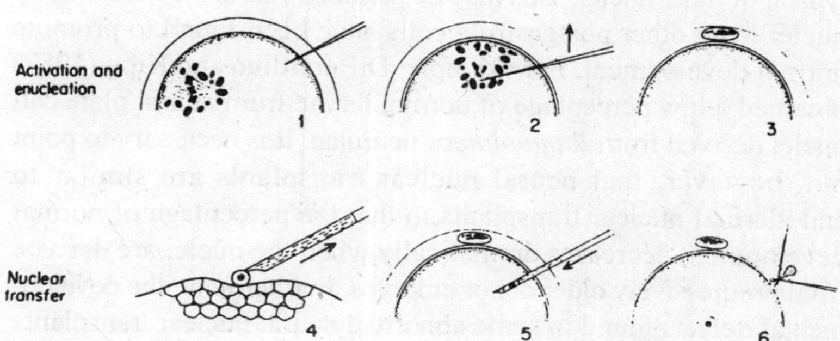

Figure 3.2 : Nuclear transplantation technique of Briggs and King. 1. Eggs are activated with a clean glass needle. 2. Activated eggs rotate and are enucleated with a second glass needle. 3. Exovate containing the nucleus is trapped by the vitelline envelope surrounding the egg. 4. Donor cells are dissociated. An individual cell is drawn into a micropipette of slightly smaller diameter than the cell. bursting it and releasing the nucleus. 5. Donor nucleus is inserted into enuncleated egg. 6. Leakage is minimized by severing the connection that forms between the egg surfae and teh vitelline envelope.

stages showed a definite restriction in developmental potential, as indicated by a dramatic reduction in the number of nuclear transplant recipients that developed normally. Transplants with nuclei from postgastrula stages exhibited a progressive loss of the ability to develop past the gastrula stage. In those transplants receiving late gastrula and postgastrula endodermal cell nuclei that did develop further, the most pronounced deficiencies were found in the size and extent of differentiation of ectodermal and mesodermal derivatives, whereas endodermal derivatives were hardly affected. This condition is termed the "endoderm syndrome." Tailbud stage endoderm nuclei were incapable of supporting normal development. These data suggest a progressive restriction in the ability of endodermal nuclei to promote normal development and further indicate that the nuclei have become irreversibly specialized as "endodermal" nuclei.

The early Briggs and King experiments with postgastrula nuclei were conducted with nuclei from only one type of cell of one species, endoderm of *Rana pipiens.* Perhaps some peculiarity of these endoderm nuclei causes this restriction, which may not be typical of other nuclei. This may be partially true, since occasionally nuclei from other postgastrula cells have been found to promote normal development. For example, DiBerardino and King (1987) obtained a low percentage of normal larvae from neural plate cell nuclei derived from *Rana pipiens* neurulae. It is necessary to point out, however, that neural nuclear transplants are similar to endodermal nuclear transplants in that the percentage of normal development decreases dramatically when the nuclei are derived from progressively older donor embryos. Furthermore, the developmental defects found in some abnormal neural nuclear transplants reflect the origin of the nuclei, since the extent of ectodermal development exceeds the development of endodermal and mesodermal derivatives. This "ectoderm syndrome" is the exact opposite of the pattern seen in transplants with nuclei of endodermal origin.

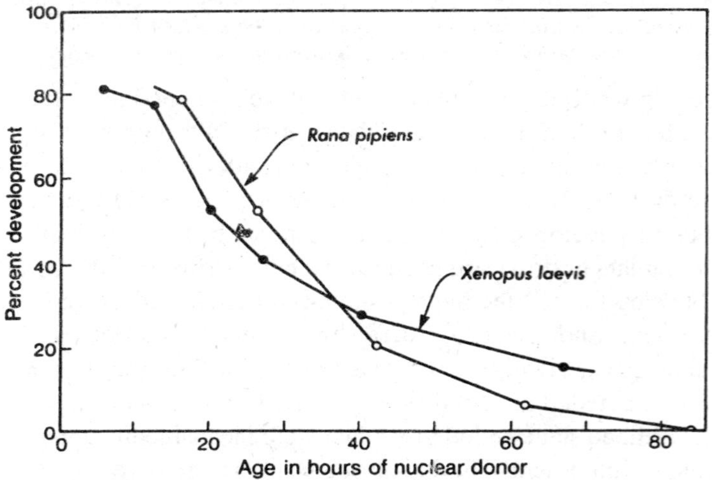

Figure 3.3 : A comparison of the capacities of* Rana *and* Xenopus *nuclei to promote normal development as the ages of donor nuclei increase.

The results of nuclear transplantation experiments with embryonic nuclei of a number of amphibian species have confirmed that as cells become more highly differentiated, their nuclei progressively become more restricted in their ability to promote development of recipient eggs. The success of nuclear transplantation that was obtained with *Rana pipiens* and *Xenopus laevis,* the South African clawed frog, as age of the donor embryo increases. The tendency for loss of developmental potential with age is similar in both species. Nuclear transplantation experiments with nonamphibian species are less extensive. Recently, success has been reported in examining the developmental potential of mammalian embryonic nuclei. Illmensee and Hoppe (1981) have indicated success in transplanting nuclei of mouse embryonic cells into mouse eggs and ob tained live-born nuclear transplant recipients.

Early embryos are removed from a donor parent. One of two mouse strains is used as a donor. Both strains contain genes for identifiable coat colors. The LT/Sv strain has a grey coat, and the CBA/H-T6 strain has an agouti coat. In addition, the LT/Sv strain carries a gene for a variant of the enzyme glucose phosphate isomerase, which can be used to identify biochemically cells containing LT/Sv nuclei. The other strain, CBA/H-T6, has an unusual chromosome constitution (karyotype) that enables these nuclei to be identified by microscopic examination.

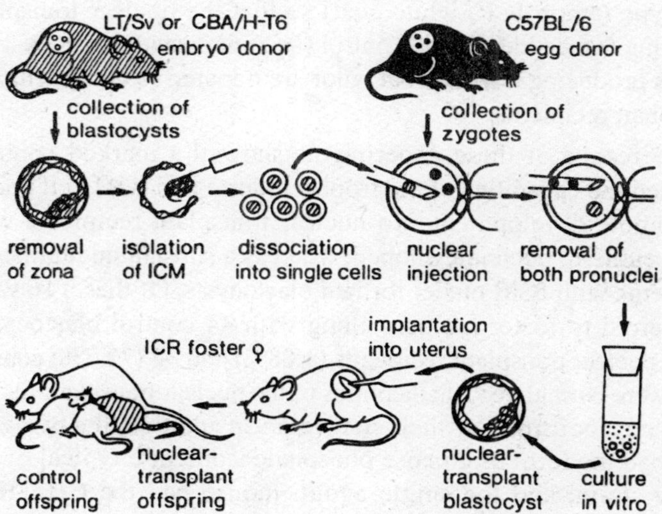

Figure 2.4 : Procedure for nuclear transplantation in the mouse.

The 4-day-old embryos used in these experiments are at the blastocyst stage. The blastocyst is surrounded by an acellular layer, the zona pellucida, and the blastocyst itself contains two recognizable regions-an outer layer of trophectoderm cells and an inner cluster called the inner cell mass (ICM). Trophectoderm cells contribute to the formation of the placenta, whereas the inner cell mass gives rise to the embryo proper. Trophectoderm and ICM are separated, and the cells whose nuclei are to be transplanted are dissociated into single cells. As with amphibian nuclear transplantation, the donor cell is broken by the transplantation micropipette. The liberated nucleus is then injected into a fertilized egg. The male and female pronuclei, which have not yet formed the zygote nucleus, are removed with the same pipette. The eggs are obtained from another mouse strain, C57BL/6. The coat color (black) and either the glucose phosphate isomerase or the karyotype can be compared to those of the donor nuclei as proof of successful removal of host nuclei and replacement by the transplanted nucleus. The nuclear transplant recipients are cultured until the blastocyst stage, at which time they are implanted into the uterus of a foster mother. The foster mother is of yet another genotype (strain ICR, white coat) so that the nuclear transplant offspring can be identified. Control (i.e., uninjected) blastocysts of strains producing a white coat color are transferred along with the transplant recipients.

The results of these experiments showed a marked contrast between the capacities of the trophectoderm and ICM cell nuclei to promote development. No nuclear transplant recipients with trophectoderm nuclei developed, whereas a substantial number of recipients with ICM nuclei formed blastocysts. Of these, 16 were transferred to foster mothers along with 44 control blastocysts. Three nuclear transplant recipients (18.8%) and 32 (72.7%) control mice were born alive. The identities of the nuclear transplant recipients were confirmed by their coat color. In addition, the two grey mice had the form of glucose phosphate isomerase typical of the LT/Sv strain, and the single agouti mouse had the CBA/HT6 karyotype. Unfortunately, the agouti mouse died before reaching sexual maturity, but both grey mice have successfully mated.

Nuclei of blastocyst ICM cells are apparently developmentally totipotent. However, we must not lose sight of two important aspects of these experiments. First, the trophectoderm cell nuclei lack potency. Hence, these nuclei have undergone restrictions during development. Second, the blastocyst stage is very early (roughly comparable to the amphibian blastula stage). Preliminary results with nuclei from cells of later stage (day 7) embryos are very intriguing. The day 7 embryo consists of several separable cell types: an outer layer of distal endoderm, an inner layer of proximal endoderm, and-still more internally-embryonic and extraembryonic ectoderm and the ectoplacental cone (a derivative of trophectoderm). Nuclei from distal endoderm, extraembryonic ectoderm, and the ectoplacental cone were unable to promote normal development of recipient eggs. On the other hand, nuclei from cells of embryonic ectoderm and proximal endoderm were capable of promoting normal development. Some of these nuclear transplant recipients have developed into live-born mice (Illmensee et al., 1981). The donor embryos in these experiments were in pregastrula stages. We await with anticipation confirmation of these preliminary experiments. It will also be interesting to learn the results of nuclear transplantation studies with still later stages to know whether nuclei in the derivatives of the embryonic ectoderm and proximal endoderm show restrictions or retain their potency.

In all species in which nuclear transplantation experiments have been done with nuclei of cells of advanced developmental stages, there is a tendency for increased developmental restrictions as donor age increases. In spite of this tendency, experiments with amphibians have shown that some nuclei of cells from postembryonic stages can promote considerable development of enucleated eggs. One series of experiments utilized nuclei that were obtained from cells of a common renal tumor, the Lucke adenocarcinoma. A small percentage of these nuclear transplants developed into tadpoles that, although not perfectly normal, did contain a large variety of functional cell types-an indication that the tumor nucleus has a large repertoire of genetic information that is potentially utilizable to direct the formation of numerous divergent cell types. Similar results have been reported by McKinnell et al. (1969), using nuclei from Lucke tumor cells derived from juvenile frogs. The latter experiments utilized a nuclear marker to ensure that the development that was obtained following nuclear transplantation was directed by the donor nucleus. The tumor cells were derived from frogs with an extra chromosome set. These triploid nuclei are readily distinguished from normal diploid nuclei, since they have a larger nuclear diameter. In addition, they contain 39, rather than 26, chromosomes. Confirmation of the triploidy of nuclear transplant individuals is an important control for these experiments, since it provides further support for the conclusion that development is directed by the tumor nucleus.

Nuclei of the germ cell line have also been transplanted to enucleated eggs. In normal development, germ cells form mature gametes that participate in zygote formation and are therefore *genetically totipotent.* This does not mean, however, that germ cell nuclei are ***developmentally totipotent*** at stages during their development. The germ cell nuclear transplantation experiments were designed to examine this question. When primordial germ cell nuclei of young tadpoles were used as donors, 40% of the complete blastulae developed into normal tadpoles (Smith, 1985). However, when spermatogonial nuclei of juvenile and adult *frogs* were transplanted, most of the complete blastulae arrested before finishing gastrulation. Of 13 complete blastulae, three developed past the gastrula stage, and one formed an abnormal larva that commenced feeding but did not survive.

Although the germ cell line retains the complete genome, germ cell nuclei apparently undergo a developmental restriction after the primordial germ cell stage, limiting the utilization of the genome for development. These experiments are important for interpreting the results of somatic cell nuclear transplantation. Clearly, genetic totipotence is not sufficient for developmental totipotence. The nucleus apparently can undergo, changes that affect the utilization of the genetic material without altering its essential nature.

The developmental potency of nuclei from postembryonic cells of *Xenopus laevis* has also been examined by extensive nuclear transplantation studies. In the initial experiments of this type,

Gurdon (1982) transplanted nuclei from Xenopus tadpole intestinal epithelium cells to enucleated *Xenopus* eggs. The techniques used for transplantation of *Xenopus* nuclei are somewhat different from those used for Rana. Instead of mechanical enucleation, the nucleus is destroyed by ultraviolet irradiation. The irradiation treatment also activates the egg to begin developing following implantation of the donor nucleus. The donor nucleus is derived from a cell obtained by chemical dissociation of tadpole intestinal epithelium. As with *Rana,* the cell containing the donor nucleus bursts as it is sucked into the transplant pipette. The liberated nucleus is then injected into the enucleated host egg. A genetic marker found in certain mutant *Xenopus* allows the investigator to determine whether the nuclei of the nuclear transplant individuals are indeed derived from the donor nucleus. Donor nuclei are obtained from a strain of *Xenopus* that has only one, rather than the normal two, nucleoli per nucleus. The host eggs are from a wild-type strain with two nucleoli. Microscopic examination of cells of the nuclear transplant individuals is conducted routinely to confirm the origin of the nucleus. A very low percentage (approximately 1.5%) of intestinal cell nuclear transplants actually developed into normal feeding tadpoles. Some of these nuclear transplant tadpoles were successfully reared in the laboratory into adult frogs.

Gurdon (1982) has argued that technical difficulties with *Xenopus* nuclear transplantation may restrict the ability of some nuclear transplant recipients to develop normally. For example, recipient embryos frequently undergo partial cleavage. Normal nuclei may be restricted to the cleaving portion of the embryo, but their potential to promote differentiation would never be realized because gastrulation would be abnormal. Therefore, the percentage of normal development may not reflect the developmental capabilities of the transplanted nuclei accurately. For this reason, serial nuclear transplantation has been conducted to maximize the opportunities for transplanted nuclei to promote normal development (Gurdon, 1962). This technique was developed earlier by King and Briggs (1956) with *Rana pipiens.*

Abnormal first-transfer *Xenopus* embryos were selected at random and dissociated. Several nuclei from each embryo were then injected singly into enucleated eggs. All serial-transfer embryos produced from a.single first-transfer embryo are a clone of genetically identical individuals, since they are all derived from nuclei that descended from a single intestinal nucleus. In every case, some of the serial-transfer embryos developed more normally than their first-transfer embryo nuclear donor, indicating that normal nuclei can be present in embryos that exhibit abnormal development. Some of the serial-transfer embryos developed into feeding tadpoles. When the results of the first-transfer generation were combined with those of the first serial-transfer generation, the percentage of intestinal cell nuclei that was capable of promoting development to the tadpole stage was raised to 7%. The most normal serialtransfer embryo was then used to provide nuclei for a second serial-transfer generation. This generation did not show any further improvement in development over that of the first serial-transfer generation. Thus, one serial-transfer generation is sufficient to demonstrate the maximal developmental capacity of the original intestinal cell nucleus. If one of the serial-transfer embryos is used subsequently as a source of nuclei for transplantation, and this process is repeated in each generation, the nuclear clone could be perpetuated indefinitely, and the developmental pattern would remain fixed throughout the lifetime of the clone.

Nuclei from another larval cell type of *Xenopus* also have been shown to be totipotent. Kobel et al. (1993) obtained a mature adult frog following serial transplantation of nuclei from epidermal cells of hatching tadpoles. Interestingly, only the nuclei from nonciliated epidermal cells promoted development; nuclei from ciliated epidermal cells, which appear to be more highly differentiated, never promoted development beyond the blastula stage.

The results of nuclear transplant experimentation have led to two generalizations:

1. As cells differentiate, their nuclei become progressively restricted in their ability to promote the development of enucleated eggs.

2. The nuclei of some differentiated cells can promote at least partial development of enucleated eggs.

These statements reflect two different points of view concerning the significance of the results of nuclear transplantation: The first emphasizes the general developmental restriction, whereas the second emphasizes the exceptions to the rule. To evaluate the relative merits of the two points of view, we should attempt to understand why developmental restriction occurs. Is it due to a normal developmental *loss* of genetic information, as proposed by Roux and Weismann? This possibility is unlikely, particularly in light of the germ cell nuclear transplants previously discussed. In those experiments, nuclei that are known to retain the complete genome nevertheless exhibit developmental restrictions during certain phases of their life cycle. Is developmental restriction due to *a modification* in the genetic material during differentiation that prevents its utilization for alternative modes of development? Finally, is there an *incompatibility* between nuclei of most differentiated cells and the cytoplasm of recipient eggs? There are no completely satisfactory answers to these questions. However, chromosomal analyses of nuclear transplants do provide some insight into the difficulty of obtaining normal development from nuclei of differentiated cells.

Numerous investigators have reported that abnormal nuclear transplants often have an aberrant karyotype; that is, the chromosome number may be modified, or the chromosomes themselves may appear abnormal when studied with the microscope. Furthermore, the severity of chromosomal abnormalities appears to be directly correlated with the extent of the developmental restriction shown by nuclear transplants. Thus, those with the most severe chromosomal abnormalities arrest earliest, whereas those with less severe aberrations can develop to larval stages. Careful studies of nuclei transplanted from late gastrula endoderm cells demonstrate that chromosomes of these nuclei frequently are unable to replicate properly when transferred to the cytoplasm of eggs. The most obvious cause of this restriction appears to be the failure of some of the chromatin to attain the proper state for replication. Chromatin exists in different degrees of condensation. Decondensation of chromatin is necessary before normal replication can occur. Variable amounts of condensed chromatin may remain in transplanted nuclei, preventing those regions from replicating. This causes both a loss of that portion of the genome and also the consequent chromosomal abnormalities that frequently are found in transplants of late gastrula and post-gastrula stage nuclei.

During normal development the mitotic rate of embryo cells is reduced considerably following the blastula stage. The continued reduction in the normal mitotic rate during development roughly parallels the decrease in the ability of nuclei from progressively later stages to promote normal development after transplantation. Furthermore, many differentiated cells never undergo mitosis. Thus, when a nucleus from such a cell is placed in the egg cytoplasm, it is in an extremely difficult situation. It is adapted to directing the specialization of one particular cell type that does not divide, but it is suddenly called upon to replicate its DNA and divide at the rate demanded during cleavage. It is not surprising that chromosomal aberrations would arise under such circumstances. Following this line of reasoning, the extent of abnormal development would depend upon the amount of damage to the genome *after* trans*plantation.*

The hypothesis that mitotic adaptability of donor nuclei affects their capacity to promote development is supported by experiments with adult *Xenopus* erythroblast and erythrocyte nuclei. Erythroblasts proliferate extensively by mitosis and are the precursors of erythrocytes, which are no longer proliferative. Thus, the effects of mitotic capacity of nuclei on their developmental potential can be tested by comparing the results of transplantation of nuclei of these two cell types. In fact, erythrocyte nuclei are incapable of promoting development beyond the early gastrula stage, whereas some erythroblast nuclei promote the development of eggs to abnormal early tadpoles. Although these nuclei differ in ways other than their mitotic activity, the correlation between mitotic activity and developmental potential is striking.

The endoderm and ectoderm syndromes previously described appear to be exceptions to the correlation between abnormal development of nuclear transplant recipients and aberrant chromosome constitution. Briggs et al. (1981), Subtelny (1985), and DiBerardino and King (1987) observed that nuclear transplant recipients displaying either the endoderm or ectoderm syndrome apparently had normal karyotypes, whereas only those with generalized developmental abnormalities could be correlated with chromosome abnormalities. These results suggest that specific nuclear differentiation occurs during embryonic development. This nuclear differentiation is stable, as demonstrated by the fact that clones can be established by serial transfer of nuclear descendants from single endoderm nuclei in which the majority of the members of the clone have the endoderm syndrome and a normal or a nearly normal karyotype. One possible explanation for these observations should be considered. Posttransplantation genetic damage that cannot be seen by microscopic examination of the chromosomes could be occurring at the gene level rather than at the gross chromosomal level, possibly as a result of the inability of specific genes to replicate properly following transplantation. This inability to replicate might be a consequence of the mechanisms that regulate gene expression . Thus, different regions of the genome may be susceptible to damage in different nuclear types.

Reversal of Nuclear Differentiation

In spite of three decades of intense investigation, nuclear transplantation has yet to provide an unequivocal answer to the question of nuclear totipotency. The experiments of Briggs and King demonstrate that nuclei do become progressively restricted in promoting development of enucleated eggs. Is it possible to reverse these changes experimentally so that differentiated cell nuclei can express their complete genetic repertoire? A number of experimental approaches have been used in attempts to reprogram nuclei of differentiated cells to function as substitute zygotic nuclei. One approach is to allow nuclei to adapt to a more rapid division rate *before* transplantation. This procedure attempts to maximize the opportunities for nuclei to make the transition to the rapid mitotic rate of cleavage and thus minimize or eliminate posttransplantation genetic damage. In these experiments, cells obtained from adult frogs are grown in tissue culture where they undergo mitosis prior to nuclear transfer.

This approach has been used with cells of the adult *Xenopus* kidney, heart, lung, and skin. Their nuclei have been shown to promote development of abnormal larvae, provided the cells are grown in tissue culture before nuclear transplantation. Most first-transfer eggs either fail to cleave or cleave abnormally, with the production of abnormal blastulae. However, if the abnormal blastulae are dissociated and used as a source of donor nuclei for serial nuclear transfer, the extent of development is increased tremendously. An extensive series of experiments have been conducted with cultured adult *Xenopus* skin cells. The differentiated state of these cells was confirmed by the detection of keratin, the protein that is specific for differentiated epidermal cells. Abnormal swimming tadpoles have been obtained in serial transfers of nuclei of these cells. Histological examination of the tadpoles has revealed that these nuclei can promote the development of a number of diverse tissues and organs, including the heart, striated muscle, brain, nerve cord, notochord, pronephros, intestine, and eye. Similar studies have shown that adult *Xenopus* lymphocyte nuclei can promote the development of abnormal swimming tadpoles.

These various experiments illustrate that mitotic preadaptation of donor nuclei can increase their developmental potential, but the recipients are'still not *completely* normal. Thus, mitotic incompatibility may not be the only problem to overcome.

Another approach has been to modify nuclei chemically to enhance their developmental potential. These experiments are based upon a natural phenomenon. As we shall discuss in detail in Chapter 6, the DNA of sperm nuclei is associated with the protein protamine. After fertilization, the egg cytoplasm promotes the remo-val of protamine from the sperm nucleus and its replacement by histones, which associate with DNA in all somatic cell types. Thus, if DNA in nuclei of cells in advanced stages of development were complexed with protamine, perhaps the egg cytoplasm could remove the protamine, which would prepare the nuclei to function as substitute zygote nuclei. In such an attempt to "remodel" chromatin, Briggs (1989) exposed Rana *pipiens* endoderm nuclei to protamine before and during transplantation and obtained a two- to threefold increase in development of recipients to larval stages. Hennen (1980) used another DNA binding molecule, spermine, in conjunction with lowered temperature when transplanting Rana *pi*piens endoderm nuclei. These conditions caused a vast increase in the ability of these nuclei to promote development to the larval stage. Whether the spermine produces its effects by actually binding to DNA has not been demonstrated experimentally. The effect of low temperature is presumably to lengthen the first cleavage cycle, which would allow the donor nuclei more time to complete DNA synthesis before the first cleavage division. These experiments demonstrate that changing the conditions of nuclear transplantation can enhance the developmental potential of nuclei. Thus, the restricted developmental potential of differentiated cell nuclei in nuclear transplantation may reflect the limited ability of egg cytoplasm to prepare these nuclei to function as unrestricted zygotic nuclei. For example, the egg cytoplasm may be more efficient at remodeling chromatin if the chromatin itself has been modified so that it resembles sperm chromatin.

Recently, experiments have been initiated to attempt nuclear reprogramming by oocyte, rather than egg, cytoplasm. The ability of oocyte cytoplasm to influence injected nuclei is firmly established. In initial experiments to test the developmental potential of nuclei transplanted to oocytes, Hoffner and DiBerardino (1980) injected tailbud stage endoderm nuclei into oocytes that were subsequently activated and had their own nuclei removed. A significant percentage (21%) of the injected oocytes formed gastrulae, and some of the gastrulae developed even further. Improvements in this technique may produce even better results and may provide the means to test the developmental potentials of adult cell nuclei that have not been amenable to transplantation to eggs.

Recent attempts to reprogram nuclei have produced promising results. However, nuclear totipotency must remain an uncertainty until some means is found to produce a high frequency of perfectly normal nuclear transplants from nuclei of adult differentiated cells. But, *at the very least,* nuclear transplantation experiments demonstrate that the nuclei of some differentiated cells can promote the differentiation of a variety of diverse cell types and are therefore pluripotent.

Totipotency in Plants

A far more satisfactory demonstration of nuclear potency comes from experiments with plant cells. The system for demonstrating totipotency was developed by F. C. Steward and evolved out of efforts during the 1940s to stimulate growth (i.e., cell division) in quiescent plant tissue. The tissue selected for these experiments was carrot root phloem. Explants of this tissue contain cells that would not normally divide again. The explants were cultured in a medium that contained coconut milk, which is the liquid endosperm of the coconut seed and normally functions to nourish and stimulate the growth of the developing coconut plant. A special culture flask was designed to allow large-scale culture of up to 100 carrot explants. The culture flasks were rotating constantly to expose the explants alternately to air and liquid.

Under these conditions the explants grew as disorganized tissue masses known as **calluses.** During rotation of the flasks, free cells were rubbed off the calluses and became suspended in the culture medium. Steward and his associates also observed small multicellular structures called **embryoids,** which resembled the normal stages in carrot embryonic development from the zygote. They interpreted the embryoids to be the products of division and differentiation of the single cells that were sloughed from the calluses. The embryoids continued developing to form small plantlets.

Presumably, then, single cells derived from carrot root phloem are totipotent, since they can be caused to divide and then recapitulate normal development. However, some controversy has arisen over the interpretation of these experiments, since the complete progression from single cell to embryoid to plantlet was not actually observed. Formal proof that. single plant cells can give rise to complete mature plants requires that a single cell be isolated from a callus and produce a complete plant. This procedure circumvents the possibility that the embryoids are actually formed from cell aggregates.

An experiment of this kind has been conducted by Vasil and Hildebrandt (1985). These investigators removed explants of stem pith from tobacco plants *(Nicotiana)* and placed the explants in tissue culture, where they formed calluses. By shaking the calluses, they dislodged large numbers of single cells and groups of cells. Single cells were removed from these suspensions and isolated to microcultures. The isolated single cells underwent cell division and formed disorganized cell masses rather than embryoids. Cell masses from the single cells were then placed on semisolid culture medium containing plant growth factors. Plantlets with roots and aerial shoots with leaves formed. These plantlets were then transferred to soil, where mature flowering plants developed.

The formation of a whole plant from a single adult somatic cell confirms that the nuclei of these cells contain all the genetic information that was originally present in the zygote. Furthermore, plant cells contain the cytoplasmic machinery that is necessary to reprogram the genome to produce a mature plant, if the proper growth factors and nutritional requirements are provided.

DIFFERENTIAL GENE EXPRESSION

Although nuclear totipotency still remains the subject of much experimental investigation, it is clear that loss of genetic information does not necessarily accompany cell differentiation. One of the most convincing examples of the retention of the genome is the polytene chromosome map of *Drosophila*. The *entire genome is* represented in certain differentiated larval cells; the band pattern (which represents the genes is identical in different cells of the same larva and even in different larvae. Molecular analyses of the genome also support the concept of genomic integrity. Genetic information is encoded in DNA in the sequence of its nucleotides. The nucleotide sequence for specific genes does not vary from cell to cell in an adult organism. There is no loss or gain of genetic information during cell differentiation. It is necessary, however, to note that there are some exceptions to this rule.

It is worthwhile at this point to consider some of the ramific-ations of the developmental integrity of the genome. The primary function of the genome is to direct the synthesis of proteins.

Theoretically, every cell has the capacity to produce the same proteins. However, the capacity to synthesize certain proteins becomes restricted to specialized cells during development. For example, red blood cells are specialized to synthesize hemoglobin, a protein that is not found in nerve cells.

Restrictions on protein synthesis must mean that certain mechanisms are operating to select portions of the genome to be utilized in different cells. Therefore, the variability in protein composition in differentiated cells must result from **differential gene expression.** The mechanisms that control development must exert their effects by controlling differential gene expression. When the developmental pathway of a cell is specified, these control mechanisms must somehow select which genes are to be expressed and which genes are to remain latent. This is the process of **cell determination.** The processes involved in utilizing those restricted portions of the genome to produce a specialized cell are cell differentiation.

Since individual cells of a developing organism are assumed to have identical genomes, determination of the pathway of differentiation to be followed in any particular cell must be imposed upon the genome from outside the nucleus. This, of course, implicates the cytoplasm in regulation of gene expression, since the nucleus is, in a sense, the captive of the cytoplasm, which completely surrounds it. E. B. Wilson referred to the cell as a. "reaction system" in which the nucleus and cytoplasm each play a role in determination of cellular characteristics. Nucleus and cytoplasm are in dynamic interaction, resulting in an altered physiological state (differentiation) in which new nucleocytoplasmic interactions come into operation.

Cytoplasmic effects on nuclei are demonstrated by experiments in which nuclei are exposed to foreign cytoplasm. Two techniques have been particularly useful for combining nuclei and cytoplasms from different cells. These are nuclear transplantation and somatic cell hybridiza**tion.** Oocytes are particularly good targets for the nuclear transplantation experiments, since they have a large amount of cytoplasm and are nondividing. Injected oocytes can be cultured in *vitro* for long periods of time, and the functions of the transplanted nuclei can then be assessed. Somatic cell hybridization is a technique for fusion of cells. If cells of different types are fused, the effects of the cytoplasm on the nuclei can be determined.

An example of nuclear transplantation to oocytes is the implantation of *Xenopus* blastula nuclei in the cytoplasm of *a Xenopus* oocyte. Oocytes are extremely active in RNA synthesis and inactive in DNA synthesis. The blastula nucleus, on the other hand, is adapted to a high rate of mitosis and therefore synthesizes DNA but little RNA. Blastula nuclei transplanted to oocyte cytoplasm undergo changes in both morphology and nucleic acid synthesis that cause them to resemble oocyte nuclei. Among the morphological changes observed, there are an enlargement of nuclear volume, dispersion of chromatin, and the appearance of nucleoli. Molecular analyses demonstrate that RNA synthesis is activated in the transplanted blastula nuclei and that DNA synthesis is inactivated.

The ability of the oocyte cytoplasm to regulate gene expression has been demonstrated in an elegant experiment by De Robertis and Gurdon (1987). Nuclei from cultured *Xenopus* kidney cells were injected into oocytes of the newt *Pleurodeles waltlii.* The proteins synthesized by the recipient oocytes were compared to those synthesized by the kidney cells and control oocytes. It was found that the recipient oocytes synthesize *Xenopus* proteins, indicating that the *Xenopus* nucleus is active in the foreign cytoplasm. However, most surprisingly, the *Xenopus* proteins include some normally synthesized by *Xenopus* oocytes, but not by cultured kidney cells. Apparently, genes that remain unexpressed in the cultured cells are expressed when the nuclei are exposed to the oocyte cytoplasm. Clearly, the oocyte cytoplasm has the ability to regulate the qualitative expression of genes.

Similar experiments have demonstrated that nuclei from adult differentiated cells can be modified by exposure to oocyte cytoplasm. Etkin (1986) injected nuclei from adult liver cells of the urodele amphibian Ambystoma *texanum* into oocytes of a related species, *Ambystoma mexicanum*. Liver cells produce the enzyme alcohol dehydrogenase (ADH), whereas oocytes do not produce ADH. Both cells produce another enzyme, lactate dehydrogenase (LDH). In the case of both enzymes, the A. texanum enzyme is distinguishable from the A. mexicanum enzyme by the technique of electrophoresis. When injected oocytes are assayed for these enzymes, it is found that they have synthesized *A. texanum* (donor) LDH but not ADH. Thus, liver cell nuclei in the oocyte cytoplasm produce proteins that are typical of the oocyte, not of the liver. This result provides further evidence for the influence of oocyte cytoplasm on nuclei.

Cytoplasmic modulation of nuclear function is also illustrated by somatic cell hybridization. The principles of animal cell fusion are similar to those of. plant protoplast fusion, although the techniques are quite different. Animal cell fusion may be facilitated by use of a virus (Sendai) that has been killed with ultraviolet light. The virus, although noninfective, binds to cell membranes and causes cells to adhere to one another. At the point of cell-to-cell contact, the membranes fuse, and the cytoplasms coalesce to form a single cell with multiple nuclei. When the cells are from different species, the hybrid cell that is formed is called a **heterokaryon.**

A particularly interesting heterokaryotic combination is between a human tissue culture cell, called a HeLa cell, and a hen erythrocyte. These cells were selected for fusion because of their contrasting patterns of DNA and RNA synthesis. HeLa cells are highly active in both DNA and RNA synthesis, whereas hen erythrocytes are inactive in DNA synthesis and synthesize very little RNA (the absolute levels of RNA synthesis in hen erythrocytes are controversial. Hen erythrocytes are particularly useful for cell fusion experiments, since the Sendai virus causes the cells to rupture, releasing the cytoplasm. When these acytoplasmic, nucleated cells fuse with a HeLa cell, all of the cytoplasm of the heterokaryon is derived from the HeLa cell.

Therefore, the effects on the erythrocyte nuclei are caused solely by the HeLa cytoplasm. After fusion the erythrocyte nuclei undergo morphological changes that resemble those seen in blastula nuclei transplanted to oocytes—they increase in volume and their chromatin disperses. These enlarged nuclei become highly active in both RNA and DNA synthesis, apparently in response to signals from the HeLa cytoplasm that regulate the level of RNA and DNA synthesis.

RNA and DNA synthesis are analyzed by use of radioactive tracers. The most specific metabolic precursor of RNA is uridine; it is not incorporated into any other macromolecule in the cell. Hence, if a source of radioactive uridine (^{3}H-uridine) is supplied to the oocyte, it will be incorporated into newly synthesized RNA. By detecting ^{3}H-uridine in nuclei, investigators can demonstrate RNA synthesis. DNA synthesis is detected by incorporation of the labeled precursor, ^{3}H-thymidine. Incorporation of labeled precursors is detected by **autoradiography.** This technique uses a photographic emulsion placed over a cytological preparation to detect radioactivity. Particles emitted from the radioisotope activate the emulsion so that silver grains are produced when the emulsion is developed by normal darkroom procedures. The silver grains are seen as tiny black dots when observed with the microscope and are located immediately over the source of radiation. RNA synthesis in the erythrocyte nuclei, which is an autoradiograph of a heterokaryon after incorporation of ^{3}H-uridine.

Cytoplasmic regulation of gene expression provides an answer to the question of how numerous cell types are differentiated from the same basic genetic information. However, it also points to an apparent enigma. Multicellular organisms begin life as a single cell, the zygote. If the cytoplasm controls gene expression, how can the derivatives of this single cell ever generate the cytoplasmic diversity that will in turn produce distinct patterns of gene expression? There are apparently two general solutions to this problem. The first is the intrinsic **heterogeneity** in the egg cytoplasm itself. If the egg cytoplasm consists of regions that differ from one another, cleavage would cause fixation of these differences, producing cells with distinct cytoplasmic constituents. The second solution involves **extrinsic influences** on the embryo. In this developmental strategy the different regions of the embryo receive positional information that establishes their fates.

Extrinsic factors are also important in regulating cell differentiation during later stages of development in both plants and animals. Formation of organs or tissues is often dependent upon developmental signals produced elsewhere in the developing organism. They could be in the form of interactions between adjacent cells, such as in embryonic induction, or they could result from long-distance interactions, such as those mediated by animal hormones or plant growth factors. However, those interactions that occur during later development are all dependent upon the initial events that establish cytoplasmic heterogeneity during early development. Since heterogeneity often results from inherent regional differences within the egg, its establishment is in turn dependent upon processes that occur during oogenesis.

The great challenge of development is to understand how it is controlled. Since the control mechanisms exert their effects by regulating gene expression, it is desirable to explain interactions of the genome with the cytoplasm, hormones, growth factors, and other regulatory factors in molecular terms. Such an endeavor requires that we have an understanding of the fundamental nature of the genetic material and its utilization by the cell. Thus, before beginning a detailed analysis of development we whall review molecular biology in the context of differentitation.

4

Genetic Information in Differentiation

THE NATURE OF GENETIC INFORMATION

The nucleus of a fertilized egg contains a vast amount of genetic information-the genome—that specifies the nature of each of the phenotypic traits, which together constitute the functional organism. Utilization of this information must be controlled so that the proper genes function at the right time and in the right cells to produce the various cells that constitute the different parts of the complete organism. The genome contains a number of different categories of information; these include ribosomal RNA genes, transfer RNA genes, and the structural genes that act as templates for messenger RNA. Additional evidence has accumulated in recent years, indicating that the genome also contains regions that do not fit the classical definition of genes. Some of these regions do not code for RNA, whereas others code for RNA that apparently does not function in protein synthesis.

The utilization of the genome for protein synthesis is a multi-step process that begins with selection of the appropriate regions of the genome, their transcription into RNA, the processing of RNA, and the translation of mRNA into protein. Investigation of the mechanisms involved in the control of information flow from the genome constitutes one of the major thrusts of contemporary developmental biology, and this chapter lays the foundation for understanding these analyses by describing the physical basis for the genome and the processes of utilization of genomic information.

THE STRUCTURE OF CHROMATIN

The DNA of multicellular organisms is not simply a free double helix within the nucleus. Instead, it is combined with proteins and RNA to form a complex called chromatin, which in turn is organized into many individual chromosomes containing compacted DNA. The condensation of DNA makes the long strands manageable within the cell nuclei. Human chromosomes, for example, contain enough DNA to form structures that would vary from 1.4 to 7.3 cm in length if the DNA were not condensed.

The degree of compaction of DNA is dependent upon the functional state of the cell. During mitosis the chromosomes are highly compact and form convenient packages that facilitate transmission of the genome to daughter cells. The most compact state of DNA occurs during metaphase, when the lengths of the chromosomes range from 2 to 10 μm. When the chromatin is in this form, it is inactive in synthetic activity. Between divisions (i.e., during interphase) chromosomes are partially unwound, forming a loose fibrous chromatin network. It is in this configuration that the bulk of transcription and replication of the genome occurs, and thus, this configuration is the most significant to development. At the completion of interphase, chromatin condenses to form recognizable chromosomes. Following mitosis, chromosomes once again unravel, and fibrous chromatin is formed.

Each round of this continuous process is designated the **cell cycle**, which is the interval between the completion of one cell division and the completion of the next division. The cell cycle is represented diagrammatically. In most cells, interphase occupies most of the cell cycle. In actively dividing cells, interphase is further subdivided into three phases-G_1, S, and G_2. The S phase is the period of DNA synthesis and is bracketed between the two G (for gap) phases. Differentiated cells may withdraw from the cycle and not undergo mitosis. Since their function is to produce specialized proteins, their nuclei usually have diffuse chromatin, which facilitates RNA synthesis.

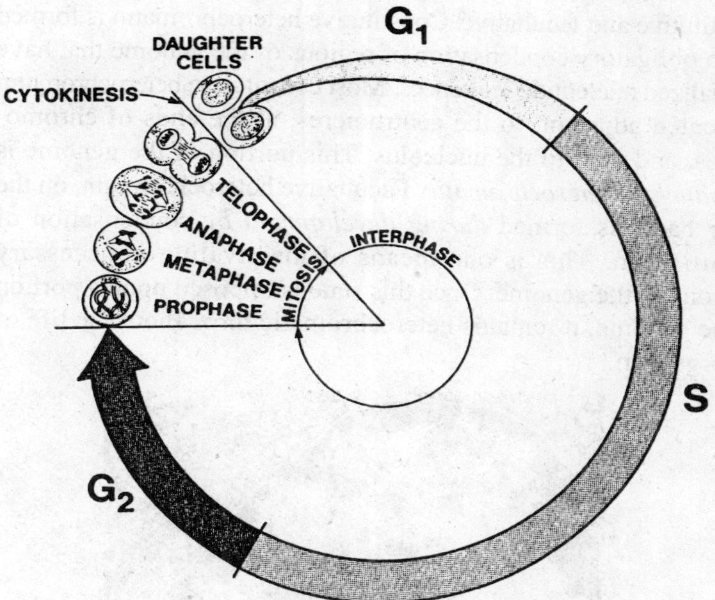

Figure 4.1 : The Cell Cycle.

Not all chromatin behaves according to the idealized pattern we have outlined. A portion of it remains condensed during interphase. It is called heterochromatin, in contrast to euchromatin, which refers to the part that undergoes the normal decondensation during interphase. These two states of chromatin have a different appearance in the interphase nucleus. Most of the central region of the nucleus is filled with diffuse euchromatin, whereas compact heterochromatin is prominent near the nuclear envelope and in association with the nucleolus.

Cytologists have long proposed that heterochromatin is inactive in RNA synthesis during interphase. This concept of heterochromatin has been confirmed by autoradiography. ^{3}H-uridine is primarily incorporated into RNA in euchromatin, whereas the genetically inactive heterochromatin shows little evidence of RNA synthesis. Since heterochromatin remains condensed when most DNA synthesis occurs, it is incapable of replicating at that time. Instead, it briefly decondenses during the latter part of the S phase and undergoes delayed replication.

Cytologists generally recognize two kinds of heterochromatin-constitutive and facultative. Constitutive heterochromatin is formed by an obligatory condensation of regions of the genome that have specialized nucleotide sequences. Most constitutive heterochromatin is located adjacent, to the centromeres, at the ends of chromosomes, and next to the nucleolus. This portion of the genome is *permanently heterochromatic.* Facultative heterochromatin, on the other hand, is formed *during development by* condensation of euchromatin. This is one means of inactivating unnecessary portions of the genome. Once this state is imposed upon a portion of the genome, it remains heterochromatic throughout the life of the organism.

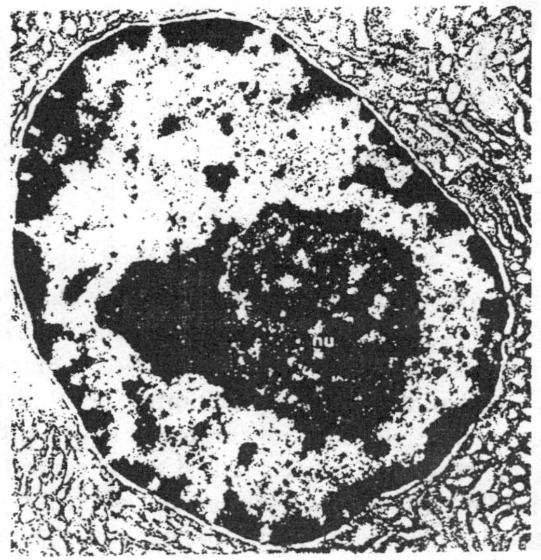

Figure 4.2 : Electron micrograph of the nucleus from a mouse pancreatic cell. The euchromatin is present as diffuse fibers occupying the interior of the nucleus. The condensed heterochromatin is associated with the nucleolus (nu) and the inner surface of the nuclear envelope (ne). × 7800.

The physical properties of chromatin are determined by both the nature of the DNA and the kinds of proteins with which the DNA is combined. Proteins are integral chromatin constituents that affect not only the gross appearance of chromatin, such as the extent of condensation, but also subtle characteristics that influence the function of particular nucleotide sequences. There are two general classes of protein constituents in chromatin: histones and non-histone chromosomal proteins (NHC proteins). The histones are positively charged molecules with a high proportion of basic amino acids, such as lysine and arginine. The percent composition of the major amino acids of the five histones of calf thymus tissue is shown in Table.

TABLE 4.1 : RELATIVE AMOUNTS OF THE MAJOR AMINO ACIDS IN CALF THYMUS HISTORIES

Histone	Major amino acids
Hl	28.7% lysine
H2A	12.5% lysine+ 13.2% alanine + 10.1% leucine
H2B	16.7% lysine+ 10.2% alanine + 10.9% serine
H3	13.6% arginine+ 13.5% alanine + 10.2% glutamic acid
H4	9.8% lysine+ 13.9% arginine + 15.9% glycine

The five histone classes differ primarily in the relative amounts of lysine and arginine; histone H1 is lysine-rich histone, 1-12A and H2B are moderately lysine-rich histones, and H3 and H4 are arginine-rich histones. These five histones are nearly universal among eukaryotes. The association of the histones with DNA appears to account to a large extent for the basic structure of chromatin. This structure is revealed in electron micrographs, which show chromatin as a series of repeating subunits resembling a "string of beads". The "beads" are spherical structures called nucleosomes. They are approximately 10 nm in diameter, and the "string" is a thin 1.5- to 2.5-nm filament of DNA. When nucleosomes are treated with the proteolytic enzyme trypsin, their structure unfolds, indicating that they are held together by proteins. When the proteins of the beaded subunits are analyzed, it is found that they are composed of an octamer of histones with two molecules each of histones H2A, 1-1213, H3, and H4.

One of the most valuable techniques for clarifying the relationship of DNA to the histones is treatment of chromatin with certain nuclease enzymes, such as micrococcal nuclease. With brief digestion, DNA fragments are produced that are approximately

200 base pairs in length. With further digestion, the enzyme cleaves the DNA fragments from their ends and progressively reduces their length. A transitory impediment to digestion is reached at 166 base pairs. This particle is subject to further degradation until a length of 146 base pairs is reached. The products of these digestion stages differ not only in DNA length but in their histone composition. The 200 base pair and 166 base pair fragments retain the octamer of histones as described previously, plus one molecule of the fifth histone, H1. However, digestion beyond the 166 base pair length releases histone H1 from the DNA (Noll and Kornberg, 1977). The release of histone H 1 upon digestion of the 166 base pair intermediate particle suggests that histone H1 is associated with the DNA that is released during digestion. We shall return subseq-uently to the relationship between histone H1 and DNA.

The 200 base pair-long fragment is a complete nucleosome. The DNA that is lost during digestion is called the linker, which is the strand that joins adjacent "beads"; it can vary in length among different cell types and between species. The set of eight histone molecules and its associated 146 base pairs of DNA compose the core particle. The length of DNA associated with core particles does not vary significantly among species. Experiments designed to determine the relationship between the DNA and the histone octamers have demonstrated that the continuous thread of DNA is periodically coiled into a helix that wraps around the *outside* of the octamers. Since the DNA molecule itself is a helix, this coiled configuration of DNA is called a superhelix. The DNA makes 1^{3}/4 turns around the protein particle. The histone octamer itself is a wedge-shaped plate; the helical path of the DNA follows surface contours on the particle. Thus, the octamer determines the architecture of the nucleosome. The nucleosomes form a repeat pattern in chromatin, with linkers alternating with cores. This alternating pattern gives chromatin the string-and-bead appearance.

The intermediate-sized 166 base pair fragments provide clues to the role of histone H1 in chromatin. These fragments contain two full superhelical turns of DNA, slightly longer than the 1¾ turns in the core particle itself. Thus, the two-turn superhelix is completed in the region flanking the core particle. The DNA that completes the two turns constitutes two 10 base pair-long strands that project from each side of the core particle. Histone H1 is associated with both these strands, binding them at a common point. Examination of the model reveals that the presence of histone H1 compacts the DNA. DNA would be extended in the absence of histone H1. According to this model, the role of histone H1 is to compact DNA. Evidence in favor of this function for histone H1 is provided by electron microscopic examination of chromatin with and without histone H1. In the presence of histone H1, chromatin is in a compact configuration in which neighboring nucleosomes are close together. In the absence of histone H1, chromatin assumes the extended "beads on a string" appearance with long internucleosome stretches of DNA.

The 10-nm chromatin fibers we have discussed so far are not seen in electron micrographs of sectioned nuclei. Instead, the micrographs reveal 20- to 30-nm chromatin fibers. Thus, the 20- to 30-nm fibers-and not the 10-nm fibers-appear to represent the native configuration of chromatin in nuclei. The larger diameter fibers are now known to be formed by the folding of the 10-nm configuration. This folding can be simulated *in vitro* by subjecting 10-nm fibers to high ionic strength solution. It is dependent upon the presence of histone H1 and is thought to occur by the polymerization of histone H1 molecules that are associated with nucleosomes, which cross-link the nucleosomes and thus form the higher-order structure.

The extent of nucleosome compaction undoubtedly varies during the cell cycle and is highest during metaphase and lowest during interphase. Furthermore, interphase chromatin probably has variable regions of nucleosome compaction, with heterochromatin being more tightly compacted than euchromatin. The regulation of nucleosome compaction, therefore, must play an important role in modulating chromatin function. Perhaps histone modification or the nonhistone chromosomal protein content of chromatin is involved in this regulation.

The functional implications of the nucleosome configuration are not well understood. However, since the nucleosome is the basic structural unit of chromatin, the processes of replication and transcription must be affected by nucleosomes, and any models of chromatin function must be compatible with this structural organization. In fact, the nucleosome structure of transcribed chromatin is modified as compared to transcriptionally inactive chromatin.

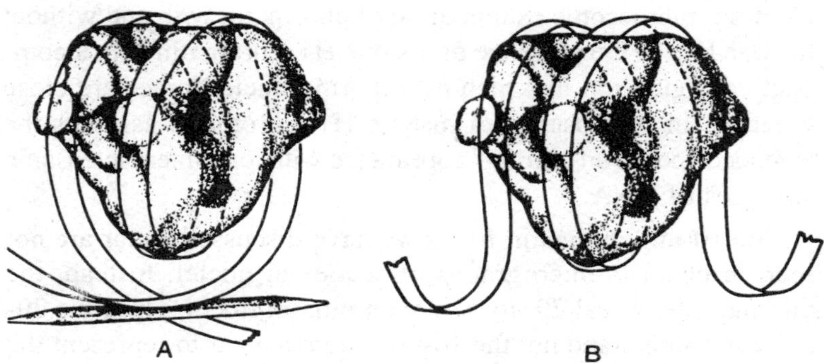

Figure 4.3 : The relationship of histone H1 to the nucleosome. A, In the presence of histone H1, DNA completes a two-turn superhelix around the wedge-shaped histone octamer. Histone H1 appears to link the strands projecting from each end of the core particle, thus sealing the two-turn superhelix. B, In the absence of histone H1, the superhelix completes only 1¾ turns, and DNA is extended from the core particle. The histone octamers represented here are based upon the model of Klug et al. (1980). The actual structure of histone HI is not known. However, it has been drawn here as an elongate structure with two free ends. These ends can interact with other histone H1 molecules to cross-link nucleosomes and, thus, compact the chromatin.

The other major class of proteins in chromatin is the nonhistone chromosomal (NHC) proteins, which fall into three categories: (1) **enzymes** such as RNA polymerase, DNA polymerase, and nucleases; (2) **structural proteins** that may modify the packaging of chromatin; and (3) **regulatory proteins** that affect the expression of specific genes. The NHC proteins are a more heterogeneous group of molecules than the histones, and thus, their characterization is more difficult. Some of the NHC proteins appear to vary from species to species and even from tissue to tissue. These variable proteins could be involved in regulating the expression of genes. NHC proteins also differ from histones in their rapid rate of turnover. Whereas histones remain associated with DNA for long periods of time with a low turnover rate, the NHC proteins are in a state of flux. Modulations in the relative amounts of specific NHC proteins will in turn modify the composition of chromatin.

The RNA associated with chromatin falls into two categories. One is **nascent RNA.** This minor chromatin constituent consists primarily of transcripts prior to their release from DNA templates. The second category includes so-called **small nuclear RNAs (snRNAs),** some of which associate with chromatin during prophase and anaphase of mitosis.

ORGANIZATION OF DNA IN CHROMATIN

A great deal of effort has been expended by molecular biologists in recent years in formulating detailed characterizations of the eukaryotic genome. This rather tedious chore is considered by most molecular biologists to be a necessary first step in understanding the function of the eukaryotic genome. This expectation results in large part from the findings that regulatory control of bacterial and bacteriophage genomes is facilitated by the structure of the genome itself. For example, genes for proteins that function together may be organized in a cluster. Regulatory control of transcription in some gene clusters is mediated by specialized nucleotide sequences adjacent to them. Although certain differences in the basic structure and function of prokaryotic and eukaryotic genomes indicate that the regulatory process is different in eukaryotes, the search for regulatory control elements within the genome remains a prime objective of contemporary molecular biology.

When the genomes of eukaryotic organisms are compared to those of prokaryotes, one of the most obvious differences is the incredible size discrepancy. For example, the human haploid genome consists of approximately 3.3×10^9 base pairs, compared to 4.2×10^6 base pairs in the bacterium *Escherichia coli.* This difference of nearly 1000 times is generally considered to be much in excess of the actual difference in informational content that is necessary to produce even a complex organism such as man. A conservative estimate of the number of genes in the human genome is 50,000. However, there is enough DNA in the genome to form almost two million genes. Apparently, eukaryotes contain a vast amount of DNA that does not fit the classic concept of a gene. This conclusion is also borne out by comparisons of the amount of DNA among eukaryotes.

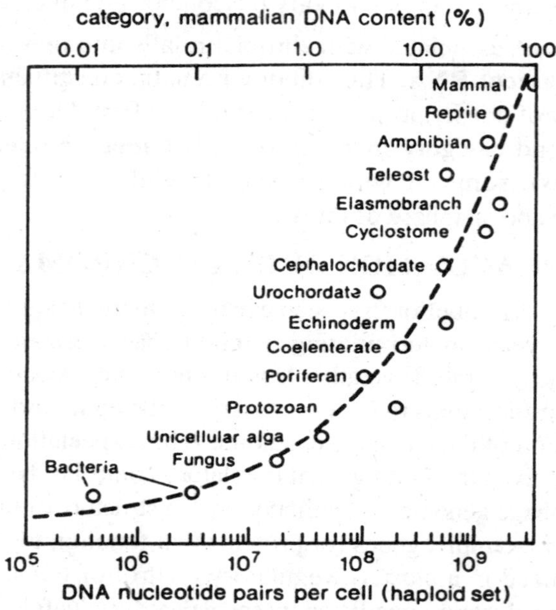

Figure 4.4 : Minimum content of haploid genome in each class of species. DNA content increases with complexity, but other members of each class may have appreciably greater DNA contents.

It is generally believed that evolution is accompanied by an increase in genetic complexity; that is, more complexity is necessary to produce a man than an amoeba. If the minimum haploid amounts of DNA (usually referred to as **C-values)** found in each taxonomic group are compared, there is a good correlation between evolutionary complexity and genome size. However, *within any group* there is a tremendous amount of variation in C-values. Most classes show a variation in genome size of approximately tenfold. The Amphibia have particularly variable amounts of DNA, with a nearly hundredfold difference. Some amphibia have as much as 25 times more DNA than man. There is no reason to assume that 25 times more genes are necessary to produce an amphibian than a man. The extra DNA is presumably not functioning as genes in the classical sense. The eukaryotic genome is obviously highly complex, with several kinds of DNA.

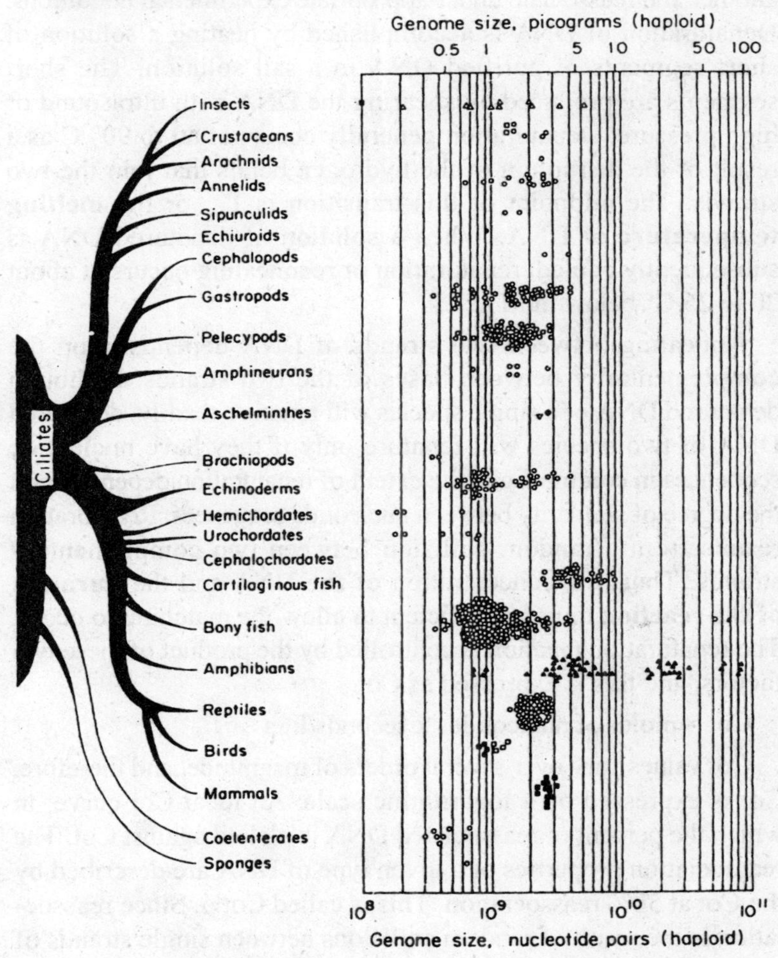

Figure 4.5 : Distribution of genome size in animals.

The basic organization of eukaryotic genomes has been determined largely by the technique of DNA reassociation, reannealing, or renaturation. This technique takes advantage of the complementary relationships between nucleotide pairs on the two strands of the DNA double helix. When the strands are separated **(denatured)** by experimental means, they will recognize one another and reassociate under appropriate experimental conditions. Denaturation of DNA is accomplished by heating a solution of short segments of purified DNA in a salt solution. The short segments are produced by shearing the DNA with ultrasound or high pressure. Denaturation generally occurs at 80 to 90° C as a result of the disruption of the hydrogen bonds that join the two strands. The midpoint of this transition is T_m, or the **melting temperature** of DNA. When a solution of denatured DNA is subsequently cooled, renaturation or reannealing occurs at about 20 to 25 C° below the T_m.

Annealing between two strands of DNA depends upon the complementarity between bases of the two strands. Although denatured DNA of a single species will renature readily, denatured DNA of two species will renature only if they have nucleotide sequences in common, and the extent of renaturation depends upon the degree of similarity between nucleotide sequences. Renaturation results from a random collision between two complementary strands. Thus, the concentration **of the DNA** and the **duration of the reaction** must be sufficient to allow the matching to occur. The renaturation reaction is controlled by the product of these two factors, and this is expressed as Cot.

Cot = moles of nucleotides x seconds/liter

Cot values vary over several orders of magnitude, and therefore, Cot is expressed by a logarithmic scale. An ideal Cot curve, in which the percent of reassociated DNA is plotted against Cot. The reassociation properties of a given type of DNA are described by the Cot at 50% reassociation. This is called Cot½. Since reassociation is the result of random collisions between single strands of DNA, the reaction rate is inversely proportional to the *number* of different sequences present. Thus, the larger the size of the genome, the more difficult it is for complementary strands to find one another, since each sequence is diluted by all the others. Renaturation will therefore occur more readily with a small genome than with a large genome. For example, the *E. coli* genome contains 4.2×10^6 nucleotide pairs, and the MS-2 virus genome contains 4×10^3 nucleotide pairs. Therefore, the concentration of any particular sequence is reduced a thousandfold in *E. coli* as compared to MS-2, requiring a Cot½ approximately 10^3 greater.

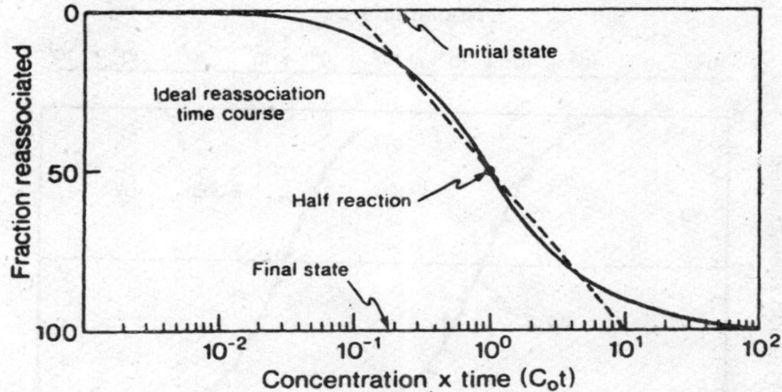

Figure 4.6 : Ideal Cot curve. The percent reassociation is plotted against the product of total concentration and time on a logarithmic scale.

Cot½ also indicates the combined length of different sequences present in a genome. This measure is called the **complexity** of the genome, which is usually expressed in nucleotide pairs. For example, if a genome consists of sequences A, B, and C, the complexity is the sum of the different sequences present; that is, A + B + C. If each of the preceding sequences were 10^2 base pairs in length, the complexity would be 3×10^2. For *E. coli* every sequence is represented once in the genome. Therefore, its complexity is identical with its genome size; that is, 4.2×10^6 nucleotide pairs.

When reassociation of two genomes is compared, each should have a Cot½ proportional to its complexity. *E. coli* is usually the standard used for such comparisons. Thus, the complexity of MS-2 (4×10^3 nucleotide pairs) is considerably less than that of *E. coli.* However, such straightforward comparisons are not possible with advanced eukaryotic organisms. We shall use the mammalian genome as an example. Since, as we discussed previously, the mammalian genome is considerably larger than that of *E. coli,* it should be more complex and have a Cot½ commensurate with its complexity. For example, the calf haploid genome contains 3.2 ×

10^9 nucleotide pairs. Since the Cot½ of *E. coli* DNA is approximately 10, the Cot½ of calf DNA should be approximately 10,000. However, eukaryotic DNA reanneals in a more complicated way than does prokaryotic DNA.

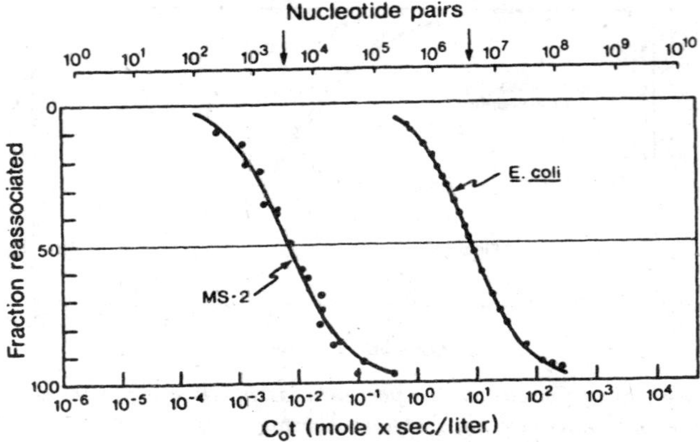

***Figure 4.7 : Reassociation of double stranded nucleic acids from MS-2 virus and* E. coli.**

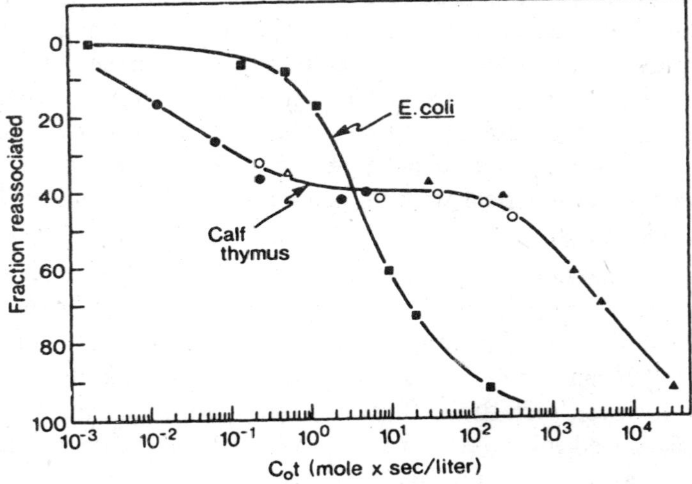

Figure 4.8 : Reassociation of calf thymus DNA compared to* E. coli *DNA.

A portion of calf thymus DNA is very rapidly reassociating, with a Cot½ of 0.03 and low complexity. This rapid renaturation indicates that a portion of the DNA is present in numerous copies. Repetition of nucleotide sequences increases their concentration and consequently causes more rapid reannealing. There is a clear separation between this DNA and a slow annealing, high complexity fraction. The latter fraction has a Cot½ of 3×10^3, which is close to the predicted value for calf DNA present in single copies. Since the reassociation of the rapidly renaturing DNA is 100,000 times as rapid as the single copy DNA, the nucleotide sequences in the repetitive DNA are repeated on the average of 100,000 times. There is an extremely broad spectrum of sequence repetition in eukaryotic DNA. This is illustrated very clearly by the Cot curve for mouse DNA. Approximately 70% of the DNA has a Cot½ value corresponding to high complexity, unique sequence DNA. An intermediate component, which corresponds to about 15% of the DNA, apparently contains sequences repeated 1,000 to 100,000 times, whereas a rapidly annealing component that constitutes 10% of the genome seems to consist of sequences repeated 1,000,000 times.

The presence of both single copy DNA and sequences with variable levels of repetition is a universal property of eukaryotic genomes-ineluding both plants and animals. There is considerable variation among organisms as to the relative amounts of the different categories of DNA; that is, single copy, moderately repetitive, and highly repetitive. However, there are some generalizations that can be made about the functions that these portions of the genome perform in cell metabolism.

The high complexity-or single copy-DNA contains most of the structural genes; that is, the genes that code for messenger RNA. Although it is unclear whether all unique sequences are structural genes, it can be said with a great deal of certainty that most structural genes are present as single copies. This conclusion has important repercussions for regulation of gene expression, since it means that if a cell becomes specialized for the production of a protein, it can rely upon only a single template (two in a diploid cell) for production of the messenger RNA for that protein.

Moderately repetitive DNA is composed of two categories of

nucleotide sequences: (1) identifiable genes that occupy a limited number of chromosomal sites (histone genes, ribosomal RNA genes, and transfer RNA genes) and (2) sequences that do not fit the classic concept of genes and are scattered throughout the genome. Highly repetitive DNA consists of short, repetitive nucleotide sequences that do not code for protein. This category of DNA is usually called satellite DNA. We shall discuss these various functional categories subsequently.

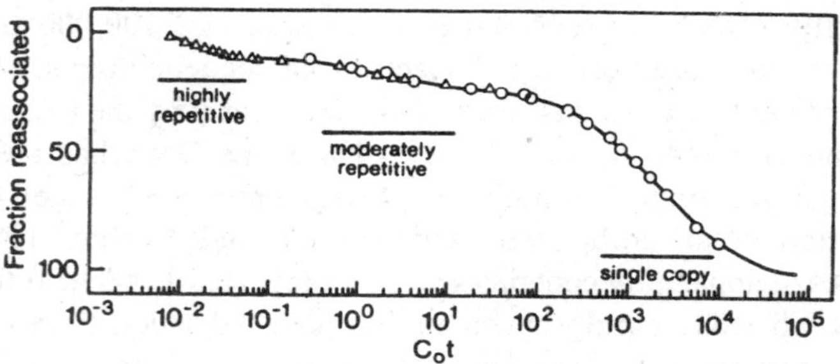

Figure 4.9 : Reassociation of mouse DNA.

Structural Genes

One of the spectacular achievements of molecular biology is the development of techniques for characterization of structural genes, most of which are present only once in the haploid genome. A unique sequence is like the proverbial needle in a haystack. It may be as short as 1000 nucleotides long embedded in a genome containing more than 10^9 nucleotides. The information in the molecular biology literature on structural genes has mushroomed since the early 1970s with the development of techniques for preparing molecular probes to locate and quantify genes and for isolating and analyzing the genes themselves. Molecular probes are molecules that can seek out and identify a specific nucleotide sequence from DNA. One of the most common probes is messenger RNA (mRNA). Since mRNA is produced as a complementary copy of DNA, it can be annealed with single-stranded DNA to form an RNA-DNA hybrid. The formation of hybrids is dependent upon a random association of RNA-DNA fragments. Thus, like DNADNA reassociation, hybridization can be characterized by the time required to form an RNA-DNA duplex and by the concentration of the reactants to determine whether the mRNA hybridizes with unique or repetitive DNA. A substantial fraction of total mRNA has been found to hybridize almost exclusively with nonrepetitive DNA. Hence, most of the mRNA molecules consist of transcripts from this type of DNA.

Similar experiments have been done with purified specific messengers as well. Specific messengers are usually obtained from cells that are committed to the predominant synthesis of a single protein. For example, red blood cells synthesize predominantly hemoglobin protein. Consequently, the messenger RNA that is isolated from these cells is relatively pure globin mRNA. A number of specific messengers have been purified from specialized cells. Nearly all of them reassociate with unique sequences. There is one clear-cut exception, however: the genes that code for histories. Histone messenger hybridizes to DNA at a rate that is considerably faster than expected for single copy sequences. Careful studies of RNA-DNA hybridization have been done with sea urchin histone mRNA and DNA and indicate that the histone genes are repeated several hundred times.

Another powerful molecular probe is synthetic cDNA, which is synthesized *in* vitro from an mRNA template utilizing the enzyme reverse transcriptase. The cDNA is identical to that portion of the gene that is transcribed into the mRNA, and therefore, its rate of hybridization with cellular DNA can also be used to determine gene frequency. Experiments of this kind have confirmed that most structural genes are present in the genome as unique sequence DNA.

Since most structural genes are present as single copies, their analysis is difficult: They are present in an incredibly complex mixture of DNA sequences. Recombinant DNA methods have simplified this problem. A specific gene can be isolated and inserted in a plasmid vector, where it is amplified numerous times. Once a sufficient amount of material is available, the amplified gene can be subjected to enzymatic analysis and its nucleotide sequence determined. This approach has proved to be most revealing when the amplified gene is a natural (instead of synthetic) gene excised from the genome, since natural genes retain sequences that may be functionally significant but are not reflected in the nucleotide sequences of their messengers.

Analysis of natural genes has led to the surprising discovery that the protein coding regions of structural genes can be interrupted by noncoding regions. The noncoding regions are called intervening **sequences,** or **introns.** The coding regions are called **exons.** Introns and exons are both transcribed, but the introns are excised from the RNA during maturation of the transcript to messenger RNA. Such messengers are, therefore, formed by splicing together noncontiguous nucleotides.

Strong circumstantial evidence for the presence of internal noncoding regions in genes has been provided by digestion of DNA with restriction endonuclease enzymes. These enzymes cleave DNA at sites containing specific nucleotide sequences. If the double-stranded cDNA for the chicken ovalbumin gene is treated with the enzyme Hind III, the enzyme has no effect on the gene, as determined by the Southern blotting procedure. This means that there are no *Hind* III restriction sites in the cDNA nucleotide sequence. This led to the expectation that one should be able to obtain fragments of DNA containing complete ovalbumin genes by treating the chicken genome with *Hind* III. However, when this experiment was done, three fragments of DNA were found to hybridize to an ovalbumin mRNA probe. This result indicates that the genomic ovalbumin gene is fragmented by *Hind* III, which recognizes internal restriction sites and cleaves the gene. Clearly, the natural gene has internal nucleotide regions that are missing from the cDNA. Since ovalbumin cDNA is a faithful copy of the ovalbumin mRNA, we can conclude that internal regions of the gene are not represented in the messenger. Extensive analysis of the ovalbumin gene with several restriction enzymes has led to the discovery of seven introns in this gene. This means that the coding portion of the gene is in eight separate pieces.

Further evidence for the presence of introns in the ovalbumin gene was provided by electron microscopic examination of the structures formed by hybridizing ovalbumin mRNA with cloned ovalbumin genes. In this technique, the DNA duplex is partially denatured to allow a complementary RNA strand to hybridize to one DNA strand and displace a single-stranded region of DNA over a comparable distance. Hybridization of the messenger with the gene causes regions of the gene corresponding to the introns to loop out because they are lacking in the messenger. All seven introns are clearly visible in this micrograph.

Introns are now known to be present in a wide variety of eukaryotic genes. Notable exceptions to this generalization are the histone genes. These genes, which have been highly conserved in evolution and are present in large numbers in every genome studied, universally lack introns.

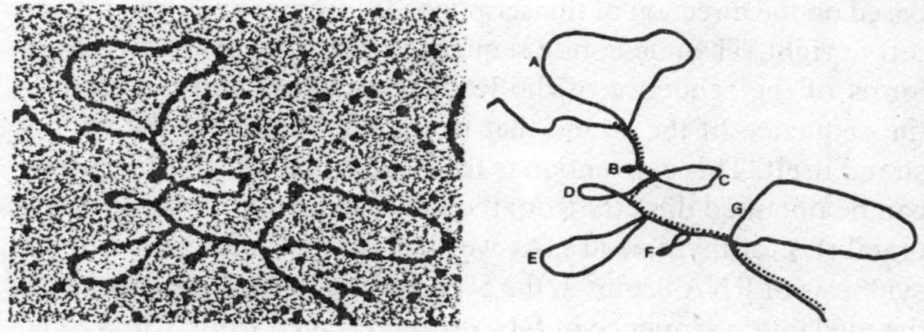

Figure 4.10 : Visualization of the seven introns of the chicken ovalbumin gene. Left Electron micrograph of the structure formed by incubation of partially denatured, cloned ovalbumin gene with ovalbumin mRNA. Seven loops are formed since the mRNA lacks complementary nucleotide sequences in those regions. Right, Interpretive drawing. The dotted line indicates the mRNA hybridized to the DNA (solid line). Loops are lettered A-G.

A more complete understanding of the organization of structural genes is provided by DNA-sequencing techniques. These techniques have been used to reveal details about both the internal organization of genes and the regions flanking the genes. Thus, introns have been precisely localized in genes, and their nucleotide sequences have been determined. Furthermore, comparisons of the noncoding regions within and adjacent to different genes have led to the discovery of so-called **consensus sequences,** which are sequences that are most frequently found in these regions. The discovery of related sequences in a wide variety of genes has led to the suspicion that they play important functional roles in gene expression. This suspicion is based upon previous discoveries that RNA codons provide signals for the initiation and termination steps of translation and that transcription initiation in prokaryotes is signaled by specific nucleotide sequences.

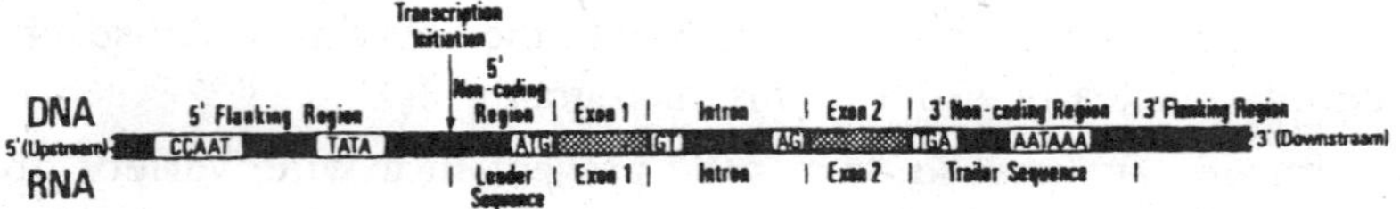

Figure 4.11 : Eukaryotic structural gene organization.

The organization of a hypothetical eukaryotic structural gene is shown in the Figure. The orientation of genes in such drawings is based on the direction of transcription. Transcription proceeds from left to right. The nucleotide sequence of DNA is always given in terms of the sequence of the RNA it produces and therefore is the sequence of the strand that is complementary to the coding strand itself. This convention is followed since the RNA sequence can be obtained directly from the DNA sequence by substituting uracil (U) for thymine (T). As we shall discuss in section 3-3, the synthesis of RNA begins at the 5' end of the RNA molecule. Thus, the nucleotide sequence of DNA is represented from 5' to 3'. The 5' flanking region, which is involved in transcription initiation, precedes the gene, and the 3' flanking region follows the gene.

The consensus sequence that is farthest upstream from the point of transcription initiation is CCAAT. Closer to the gene is the sequence TATA. The latter is usually called the Goldberg-Hogness box, after the investigators who first recognized it. The site of transcription initiation precedes the translation initiation codon ATG (AUG in the RNA). The portion of the transcript upstream from the first exon is called the **leader sequence.** The gene has two exons and one intron. Each structural gene intron begins with the dinucleotide GT and ends with AG. The final exon is followed by a translation termination codon, TGA or TAA (UGA or UAA in RNA). Transcription continues past the translation termination codon until the point of transcription termination is reached. The segment of RNA produced from the region downstream from the last exon is called the **trailer sequence.** It may contain a hexanucleotide sequence AAUAAA (AATAAA in DNA). Downstream from the transcribed portion of the gene is the 3' flanking region.

The consensus sequences we have described are found in many structural genes, but not in all of them. Variations are found in some cases. For example, the Goldberg-Hogness box in some β-globin genes is CATA, rather than TATA. Some genes lack certain consensus sequences. For example, histone genes lack the AATAAA hexanucleotide in the 3' noncoding region. We shall discuss the functions of the various consensus sequences in subsequent sections of this chapter.

Even though individual structural genes are present as single copies, many of them are accompanied by closely related sequences that are often clustered in the same chromosomal region. These closely related genes constitute **multigene** families. Families of genes probably arose during evolution by repeated duplication of ancestral genes followed by divergence of nucleotide sequences. These descendants may code for related polypeptides. Other descendants seem to be evolutionary relics with no apparent function **(pseudogenes).** Pseudogenes have acquired changes in nucleotide sequence that are so deleterious as to render them nonfunctional. The presence of pseudogenes helps to explain a portion of the apparent excess of DNA in eukaryotic genomes.

Ribosomal RNA Genes

There are four kinds of ribosomal RNA in eukaryotic cells-5S, 5.8S, 18S, and 28S rRNA. The names given to rRNA indicate the relative sizes of the molecules. The three larger molecules are transcribed in tandem from closely linked genes, with the gene for 5.8S rRNA situated between the 18S and 28S genes; the 5S RNA is transcribed from a separate, unrelated region of the genome and will be discussed later. Each complex of 18S-5.8S-28S rRNA genes ,is present in several hundred copies in most eukaryotes. Therefore, these genes are categorized as moderately repetitive genes. They are located at the nucleolar organizer (NO) locus of eukaryotes, which is the site of formation of the nucleolus. Many organisms have one NO region per haploid chromosome set and therefore two NO regions in diploid cells. This situation is seen, for example, in *Drosophila melanogaster* and in *Xenopus laevis*. There are certain exceptions, however. For example, the human karyotype has NO regions on several chromosomes. Genetic manipulation of both *Xenopus laevis* and *Drosophila melanogaster* has been used to produce organisms with variable numbers of nucleoli. The normal diploid cell in both these organisms has two nucleoli, but variants of both species can have either no nucleoli or unusual numbers of nucleoli: one or two in *Xenopus* and frcm one to four in *Drosophila*.

Hybridization of rRNA with DNA from organisms with variable numbers of nucleoli has established that the level of hybridization is proportional to the number of NO regions, indicating that the genes for rRNA are located in those regions. Another means of establishing the identity of the NO as the site of the rRNA genes is RNA-DNA in situ hybridization, whereby cytological preparations of chromosomes that have been treated with alkali to denature the DNA are incubated with radioactive rRNA. The labeled RNA will hybridize with the DNA containing complementary nucleotide sequences. Autoradiography is then used to localize the site of radioactivity, which corresponds to the site of the ribosomal genes. Following *in* situ hybridization, autoradiographs show that the rRNA genes are indeed located at the nucleolar organizer.

As previously mentioned, ribosomal DNA is organized into gene complexes, each of which contains one 18S gene, one 5.8S gene, and one 28S gene in the order listed from the 5' to the 3' end (as before, the polarity of the DNA reflects the polarity of the RNA it produces; that is, the 5' end of RNA is said to be produced from the 5' end of the gene). The multiple ribosomal gene complexes are separated from one another by **spacer DNA,** which is not transcribed. This kind of gene arrangement is called **tandem repetition.** As we shall discuss in detail in section 3-3, each gene complex is transcribed as a unit, resulting in a large precursor molecule containing the 18S, 5.8S, and 28S elements. The large molecule is later processed to yield the three smaller molecules.

The repetition of rRNA genes provides cells with multiple coding units for the efficient production of this class of RNA. During oogenesis, even this level of repetition may be insufficient to produce the required amount of rRNA, and numerous extra rRNA genes are produced by a selective replication of the nucleolar organizer. This process is called **gene amplification** and has been observed in some fish, insects, and amphibians. Apparently, amplification involves replication of the nucleolar organizer. These copies are released from the chromosomes, and the copies themselves are used as templates for the production of still more replicates. Each free NO region then forms a small nucleolus that is free within the oocyte nucleus and located just below the nuclear envelope. The amphibian oocyte nucleus may have more than 1000 of these nucleoli. The small nucleoli that are produced consist of a fibrous core surrounded by a granular cortex. The core contains the ribosomal DNA and is, therefore, the site of synthesis of ribosomal RNA.

After synthesis the rRNA detaches from the DNA and is transferred to the cortex, where it undergoes processing prior to its packaging into ribosomes. The structure of the DNA within the core is difficult to discern because the core is so compact; its structure is best studied by breaking open the nucleolus to allow the DNA to unravel. This can be accomplished by placing the nucleoli in distilled water. The nucleolar DNA is then found to be a single closed circular molecule.

Besides providing the oocyte with additional rRNA coding units for rapid synthesis of rRNA, amplification provides investigators with a convenient source of nucleoli for experimental analysis. Consequently, much of the information we have concerning organization of rRNA genes and the mechanisms of their transcription has been obtained with oocyte nucleoli.

5S and Transfer RNA Genes

These two categories of genes produce small-sized RNA. However, in addition to size, there is another reason for considering them together: These genes are organized differently from the other genes of eukaryotes, and they are transcribed by an RNA polymerase that is selective for them.

The DNA that encodes 5S and transfer RNA is moderately repetitive. Furthermore, the tRNA genes are actually a heterogeneous group of genes that encode the several kinds of transfer RNA. In Xenopus, each of the 43 different tRNA genes is repeated an average of 200 timesmaking a total of about 8600 genes. There may also be heterogeneity among the 5S RNA genes. In *Xenopus laevis,* for example, there are three kinds of genes that code for 5S RNA: somatic, oocyte, and trace oocyte. The genes that code for each kind of 5S RNA are present as multigene clusters, with units consisting of a gene plus spacer DNA tandemly repeated hundreds to thousands of times. A repeat unit may also contain a pseudogene. The oocyte 5S RNA genes are present in the highest number in the X. *laevis* genome. They are expressed only during oogenesis and are inactive in somatic cells. The oocyte 5S RNA genes are present as approximately 24,000 tandemly repeated 120 base pair coding units per haploid genome. Clusters of oocyte 5S RNA genes have been localized to the telomeres of the long arms of most, if not all, of the Xenopus chromosomes by in *situ* hybridization.

As already mentioned, the individual gene units that encode the various kinds of tRNA and 5S RNA have a novel organization that is related to their mode of transcription.

Interspersed Repetitive DNA

In addition to the well-characterized repetitive sequences previously discussed, the eukaryotic genome also contains repetitive sequences that are scattered throughout the genome. These sequences are interspersed between the unique sequences (at least some of which are structural genes). It is now thought that most, if not all, unique DNA sequences are flanked on either side by moderately repetitive DNA. The initial evidence for this genome organization was obtained from renaturation studies on relatively long fragments (20,000 nucleotides) of *Xenopus* DNA. Since unique sequences account for the majority of DNA, one would expect most of these fragments to consist of unique sequences and therefore reanneal under high Cot conditions. However, they actually reassociate under low Cot conditions that allow moderately repetitive DNA to react. This result indicates that the fragments must contain moderately repetitive sequences in addition to unique sequence DNA and led to the proposal that these moderately repetitive sequences are interspersed between unique sequences. This proposal has since been supported by a large body of evidence from studies with cloned genomic DNA. Cloned fragments containing structural genes have frequently been shown to contain repetitive DNA in the regions flanking the genes. For example, a fragment containing the rabbit R-globin gene possesses a repetitive segment downstream from the gene, which is also present at several scattered sites throughout the genome. The rabbit genome contains about 5000 copies of this particular sequence.

Recently, repetitive sequences themselves have been cloned and used as hybridization probes. In addition, the nucleotide order of some of these sequences has been determined. One result of studies with repetitive sequence probes has been the discovery of families of sequences that hybridize to the same probe. Members of a **repetitive sequence family** have high sequence homology with one another and may share consensus sequences at specific sites. For example, a single 300 nucleotide-long, repetitive sequence family predominates in the human genome. This family of approximately 300,000 members has been called the **Alu family,** because each member possesses a restriction site for the enzyme Alu I. Other mammals have recently been shown to have *Alu* family equivalents.

The function of interspersed repetitive DNA is as yet unknown. An important question is whether these sequences are transcribed and, if so, what the function of the transcripts is. Cloned sea urchin repetitive sequences have been used to probe the RNA for such transcripts. The results of these experiments indicate that the sequences are transcribed. However, the transcripts are restricted to the nucleus. Furthermore, there are tissue-specific differences in the relative frequencies of different repeat family transcripts. These two characteristics suggest that the repetitive sequence transcripts may serve some regulatory function in the nuclei, such as the regulation of structural gene expression or processing of transcripts.

In a similar experiment a cloned probe for human *Alu* family sequences has been demonstrated to hybridize with a high proportion of human messenger RNA. Thus, these sequences are transcribed and remain an integral part of the transcripts that can function as messenger RNA. The roles the Alu sequences might play in messenger RNA are unknown.

An interesting and possibly highly significant class of repetitive sequences has been identified whose members are capable of being inserted at several locations in the genome. These sequences, which are found in a wide variety of organisms, are called **transposable elements** or **mobile dispersed elements.** Their transposable nature is indicated by the fact that they occupy quite different positions in the genome among closely related strains of the same species. Thus, they are capable of excision from sites they occupy and of reinsertion at alternate sites. Although the presence of a transposable element may affect the expression of genes in its vicinity, the functional roles of these elements are poorly understood.

Satellite DNA

Rapidly reannealing DNA consists of short, tandemly repeated nucleotide sequences that do not code for protein. This category of DNA is sometimes called satellite DNA. This term refers to the fact that the highly repetitive DNA has a buoyant density that is different from that of bulk DNA when subjected to analytical density gradient centrifugation. Thus, during centrifugation the bulk DNA forms a major band, whereas the smaller amount of repetitive DNA forms a small satellite. The difference in buoyant density is due to differences in base composition between the two classes of DNA. Since the rapidly reassociating DNA consists of short sequences that are repeated approximately one million times, it is enriched for certain bases; therefore, its base composition differs greatly from the remainder of the DNA, resulting in different buoyant density.

The correspondence between satellite DNA and rapidly

reassociating DNA is demonstrated by density gradient analysis of purified rapidly reassociating DNA. If denatured DNA is renatured under very low Cot conditions, the double-stranded rapidly reassociating DNA can be separated from the remaining DNA (unique and moderately repetitive sequences) by passing the solution over a column of hydroxyapatite (calcium phosphate). The hydroxyapatite selectively binds double-stranded DNA, but single-stranded fragments pass through the column. The bound DNA can then be removed from the hydroxyapatite by raising the salt concentration of the buffer. Under high salt conditions the DNA no longer binds to the hydroxyapatite and is flushed **(eluted)** from the column. The highly repetitive DNA prepared by this technique of affinity **chromatography** has a buoyant density that is identical to that of satellite DNA.

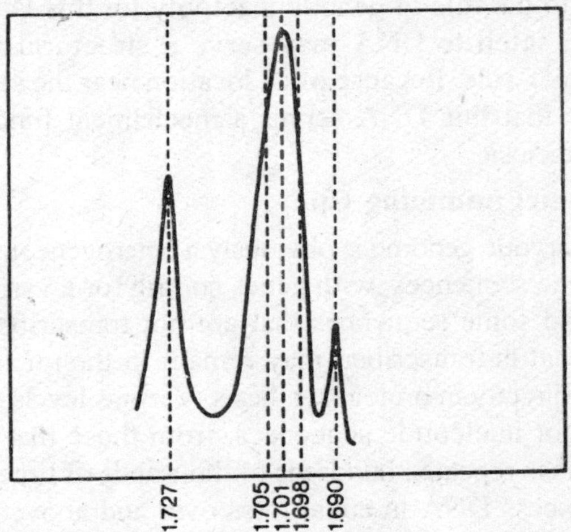

Figure 4.12 : Analytical ultracentrifugation of mouse DNA in cesium chloride. The peak at 1.727 is a density marker. Main band mouse DNA reaches equilibrium at 1.701 and a satellite peak forms at a density of 1.690.

Satellite DNA is localized to specific chromosomal regions where it is present as large blocks of tandemly repeated nucleotide sequences. These locations have been established by *in situ* hybridization. A cytological preparation of chromosomes on a microscope slide is treated with alkali to denature the DNA, and the preparation is incubated with radioactively labeled RNA that is synthesized *in vitro* from satellite DNA. Silver grains in developed autoradiographs indicate the presence of radioactive RNA that has hybridized with the chromosomal DNA. The silver grains are localized predominantly over the constitutive heterochromatin surrounding the centromeres of metaphase chromosomes. Hence, the major DNA component of constitutive heterochromatin is the rapidly reassociating satellite DNA. A variety of studies using *in situ* hybridization as well as other techniques have provided evidence that satellite sequences may also be present in nonheterochromatic regions of the genome.

It is generally assumed that satellite DNA is inactive in RNA synthesis. However, this DNA is transcribed in amphibian oocyte lampbrush chromosomes. Transcription in oocytes is an apparent exception to the rule of genetic inactivity for this DNA. In other cell types, satellite DNA may serve a structural, rather than informational, role. Because of its location near the centromere, it is possible that this DNA serves a mechanical function during mitosis or meiosis.

The Genome: Summing Up

The eukaryotic genome is obviously a heterogeneous assortment of nucleotide sequences, with genes coding for a variety of types of RNA and some sequences that are not transcribed. Some of the RNA that is transcribed may remain in the nucleus and not participate directly in protein synthesis. Various levels of repetition are found for nucleotide sequences, from those that are present singly to those repeated hundreds of thousands of times. Thus, the apparent excess DNA in eukaryotes over and above the amount needed to code for proteins has an explanation, but its significance to the phenotypes of organisms that contain it is only partially understood and remains a challenge to those concerned with the role of the genome in development.

THE UTILIZATION OF GENETIC INFORMATION DURING CELL DIFFERENTIATION

RNA Synthesis

The initial step in cellular utilization of the information stored in the genome is transcription of DNA to form RNA. The enzymes responsible for this activity are a group collectively called DNA-dependent RNA polymerases. Three distinct polymerase enzymes are responsible for transcription in eukaryotes; each functions on different classes of DNA. RNA polymerase I is found preferentially in the nucleolus and is utilized in transcription of rDNA. The two other polymerases, II and III, are described as being "nucleoplasmic"; that is, they are found in the non-nucleolar regions of the nucleus, presumably associated with chromatin. Polymerase II transcribes structural genes, whereas type III transcribes the genes for transfer RNA and 5S RNA.

The differential inhibition of polymerase function is often an important experimental manipulation in the study of cell differentiation. It is useful, for example, in determining whether *de novo* transcription is essential for the appearance of a particular differentiated trait or for the synthesis of a particular protein. The most commonly used inhibitor **is actinomycin D**, which at low concentrations inhibits rRNA synthesis and at higher concentrations inhibits synthesis of all classes of RNA. Another inhibitor, **a-amanitin,** is more specific; low concentrations selectively prevent polymerase II function, higher concentrations inhibit polymerase III, whereas polymerase I is unaffected by these concentrations of this inhibitor.

Polymerase molecules transcribe DNA in a linear fashion by sequential addition of ribonucleotides to a growing RNA chain. Each nucleotide in RNA has the same orientation, giving the chain a definite polarity, which is opposite that of the DNA template. The end terminated by the 5'-carbon atom and its associated $(POI)_3$ group is called the 5' end, whereas the end containing the 3'-carbon atom and its OH group is the 3' end.

During transcription a polymerase molecule attaches to the beginning of the region of DNA to be transcribed (a **transcriptional unit).** The site at which the polymerase initiates transcription is called the **promoter**. The promoter also influences the efficiency of initiation. Attachment of the polymerase to the DNA causes the double helix in the attachment region to unwind. The polymerase then moves along the DNA, causing local unwinding and reformation of the helix as it proceeds. Only one DNA strand is transcribed by the polymerase, which always moves in the 3'-+5' direction of this strand. This produces an RNA molecule with opposite polarity-it begins at the 5' end and terminates at the 3' end. The growing RNA strand remains attached to the polymerase during the entire transcriptional process, and when this complex reaches the 5' end of the transcriptional unit, both the polymerase and the completed transcript are released from the DNA. The polarity of a gene is given in terms of the noncoding (or antisense) strand. Thus, the beginning of a gene is said to be the 5' end, whereas the terminus is the 3' end. Unless stated otherwise, we shall follow this convention when discussing gene organization.

The structure and function of promoters is best understood in prokaryotes. The prokaryotic RNA polymerase recognizes and binds to two regions of DNA, located approximately 10 nucleotides and 35 nucleotides, respectively, upstream from the mRNA start site. The former, which is called **Pribnow's box,** has the sequence TATAATG. The segment at 35 nucleotides upstream has the consensus sequence TGTTGACAATTT, although there is considerable variation of the latter sequence among prokaryotes. These promoter regions are remarkably similar to consensus sequences found in the 5' flanking regions of eukaryotic polymerase II genes. These are the TATA, or Goldberg-Hogness, box and the CCAAT box, which is located farther upstream.

Analysis of the roles of the eukaryotic consensus sequences in transcription of polymerase II genes involves observing the effects on transcription of modification or deletion of these sequences. Genes with aberrant 5' flanking regions are cloned, and transcription of the cloned sequences is assessed by either in *vitro* transcription with polymerase II or the technique of **surrogate genetics.** The latter procedure involves introducing the cloned genes into the nucleus of another cell and allowing the endogenous RNA polymerase II of the host nucleus to transcribe them. Exogenous genetic material may either be injected into the nucleus of a living cell (usually *a Xenopus* oocyte) or introduced into tissue culture cells by transformation. Experimental analysis of the TATA box indicates that it directs the polymerase to initiate transcription about

30 nucleotides downstream from it. Thus, it specifies the RNA start site. Regions located farther upstream-among them the CCAAT box-exert an influence on the efficiency of transcription. Therefore, the eukaryotic polymerase II promoter is a multisite entity in the 5' flanking region.

The promoter of ribosomal RNA gene transcription by RNA polymerase I is also located in the 5' flanking region preceding each 18S-5.8S-28S rRNA gene complex. However, genes transcribed by RNA polymerase III (5S RNA and transfer RNA genes) are unique in that the sequences directing specific initiation of transcription are *intragenic and* are flanked by coding sequences. Thus, the polymerase III promoter specifies that transcription begin at a site far upstream from it.

One of the most powerful means of studying transcription is actually to observe chromatin in the act of RNA synthesis with either the light or electron microscope. The initial observations of transcription were light-microscopic studies of so-called giant chromosomes. One kind is the lampbrush chromosome, which appears during the diplotene stage of meiosis in a wide variety of plant and animal species. The other kind of giant chromosome is found in certain enlarged cells of the larvae of *Drosophila, Chironomus* and some other dipteran flies. These polytene chromosomes have provided considerable information concerning regulation of transcription.

A technique developed by Oscar Miller allows observation of transcription with the electron microscope. Essentially, this procedure involves a **gentle dispersion** of nuclear contents to separate constituents without breakage of chromatin or loss of the nascent RNA. This step is followed by **fixation,** the **spreading** of the chromatin on electron microscope grids, and **staining,** which allows identification of the chromatin and RNA. Ribosomal RNA synthesis is perhaps the most visually pleasing transcriptional process that has been observed with the electron microscope. The initial observations of ribosomal DNA transcription were made on chromatin prepared from amphibian oocyte nucleoli. Numerous observations have subsequently been made on nucleolar transcription in a wide variety of species, and the pattern is virtually identical wherever it is observed.

Under low magnification, nucleolar chromatin has the appearance of a series of "fuzzy" regions alternating with smooth chromatin. The fuzzy regions are called **matrix units.** At higher magnification, the matrix units are seen to consist of a sequence of progressively longer fibrils attached to chromatin by a small granule. The fibrils are nascent RNA transcripts, and the granules at the base of the fibrils are thought to be RNA polymerase I molecules. Transcription apparently begins at the narrow end of the matrix unit with the binding of a polymerase molecule to the chromatin. The polymerase then moves along the DNA, progressively adding ribonucleotides to the growing RNA chain until it reaches the end of the matrix unit and detaches from the chromatin. As soon as one polymerase has moved a sufficient distance along the DNA, another polymerase attaches, and so on. This "assembly line" process allows for the matrix DNA to be transcribed simultaneously by several polymerase molecules.

As we discussed earlier in this chapter, the 18S, 5.8S, and 28S ribosomal RNA genes are organized as gene complexes, repeated numerous times, and the gene complexes are separated from one another by nontranscribed spacer DNA. The DNA within each matrix unit is one gene complex, and the smooth chromatin represents the nontranscribed spacers. Transcription is initiated at the 5' end of the gene complex (narrow end of the matrix unit), as polymerase I binds to the 5' end of the complex-as specified by the polymerase I promoter located thereand progressively transcribes the 18S, 5.8S, and 28S genes, producing a single, large precursor molecule that contains all three kinds of ribosomal RNA. The functional 18S, 5.8S, and 28S molecules are released from the precursor by processing. Not only is each gene complex transcribed simultaneously by several polymerase molecules, but all genes complexes are transcribed at the same time. Ribosomal transcription is obviously a highly efficient process.

Transcription of nonribosomal interphase chromatin of somatic cells has been observed with the electron microscope only in recent years. The few nonribosomal transcriptional units that have been seen are similar to ribosomal transcriptional units in that they also have a series of progressively longer fibrils associated with the chromatin. However, in contrast to ribosomal transcription, the transcriptional units are found not as tandem repeats but as isolated regions of transcriptional activity. Apparently, they are unique nucleotide sequences and may represent structural genes. The nascent RNA fibrils in nonribosomal transcriptional units are farther apart than those in ribosokal units, indicating that fewer RNA molecules are synthesized simultaneously on these genes. Careful examination of these transcriptional units at high magnification reveals that the transcribed chromatin is beaded.

The presence of nucleosomes in transcriptionally active chromatin makes it necessary to account for this configuration in any explanation of the transcriptional process. How does the polymerase molecule traverse the nucleosome? Do nucleosomes in active chromatin differ from those in inactive chromatin? Although we know little about the movement of the polymerase, there is strong evidence that the chromatin of transcribed DNA has a somewhat different nucleosomal configuration than that of nontranscribed DNA.

From the moment of its synthesis, RNA is combined with protein to form a complex called **ribonucleoprotein (RNP).** In fact, the growing fibrils seen with the Miller electron microscopy technique are visible because they are stained for protein, not RNA. Following transcription, completed transcripts are released from chromatin and transported to the cytoplasm as RNP complexes. The type of complex formed depends upon the nature of the RNA. The ribosome is a complex of the four kinds of rRNA-5S, 5.8S, 18S, and 28S-with specialized ribosomal proteins. Structural gene transcripts are combined with proteins to form messenger RNP particles. Considerable evidence suggests that messenger RNA remains combined with protein throughout its functional life in the cell.

PROCESSING OF RNA

Transcription is only the first step in the sequence of events that must occur so that a gene may exert its influence on cellular function. The raw transcript is a diamond in the rough; it must undergo a number of modifications in the nucleus and then be transported to the cytoplasm, where it becomes functional. These modifications include addition, deletion, and alteration of nucleotides.

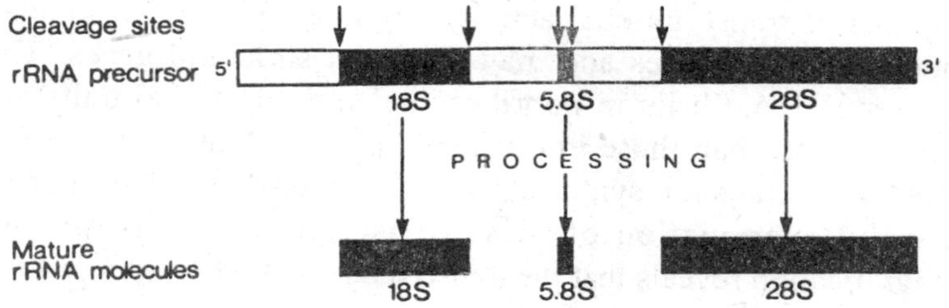

Figure 4.13 : Processing of ribosomal RNA precursor molecule. The precursor has the following regions from the 5' to 3' end: spacer- I 8S-spacer-5.8S-spacer -28S. The spacers are cleaved away, releasing the individual rRNA molecules.

The primary transcript may be longer than its cytoplasmic derivative. Enzymatic cleavage of the precursor is necessary to trim away excess nucleotides. For example, the ribosomal RNA precursor is a complex molecule containing 18S, 5.8S, and 28S segments, separated from one another by spacer regions Nulease enzymes in the nucleolus trim away the spacer regions, which eleases the 18S, 5.8S, and 28S molecules. A number of nucleotides of ribosomal RNA are also altered by the addition of methyl groups to the ribose moieties. The primary ribosomal transcript is, therefore, subject to extensive **post-transcriptional modification.**

Nonribosomal transcripts also undergo posttranscriptional processing. Like the rRNA precursor, nonribosomal transcripts in the nucleus are very large molecules. In mammalian cells, for example, nuclear transcripts vary in length from about 5000 nucleotides to greater than 50,000, whereas messenger RNA molecules in the cytoplasm range in size from 500 to 5000 nucleotides. Nuclear RNA is usually called **heterogeneous nuclear RNA,** abbreviated as **HnRNA.** It is a mixture of several molecular size classes. Most of the HnRNA is very short lived and is degraded in the nucleus, never reaching the cytoplasm . At least a portion of the RNA that is restricted to the nucleus includes transcripts of the interspersed moderately repetitive DNA.

There is a large body of evidence that at least some of the HnRNA molecules are large precursors from which mRNA molecules are derived by processing. HnRNA and mRNA have many properties in common, which suggests a possible relationship. For example, both are called DNA-like RNA because their base compositions are similar to that of the average DNA composition, an indication that both kinds of RNA are transcribed from generalized DNA. Ribosomal RNA, on the other hand, is transcribed from a limited portion of the genome with a high ratio of guanine and cytosine, which is reflected in the base composition of the RNA.

Establishing a precursor-product relationship between HnRNA and mRNA has been more difficult than determining the relationship between nucleolar RNA and ribosomal RNA. Strong evidence for this relationship is the discovery of messenger RNA sequences in HnRNA molecules with cDNA probes. For example, globin cDNA has been used to detect the high molecular weight nuclear precursors to mammalian globin mRNA and to trace the processing of the globin gene transcript. The nuclear transcript is a large 15S molecule, whereas its cytoplasmic messenger derivajive sediments at 10S. The latter is derived from the precursor by removal of excess nucleotides, modification of existing nucleotides, and posttranscriptional addition of certain nucleotides.

The processing of HnRNA from its synthesis to the formation of mRNA is now becoming clear. As we discussed previously, the beginning of the transcript (the 5' end) is marked by a triphosphate group, which is part of the initial nucleoside triphosphate at the transcription initiation site. After initiation has occurred the 5' end is modified by the addition of a guanosine nucleoside that is attached in opposite polarity to the rest of the RNA molecule. In addition to having reversed polarity, this guanosine is also unusual in that it contains a methyl group bound to the guanine portion of the molecule. This methylated guanosine is called the **5' cap.** Transcripts may also be methylated in the nucleotide adjacent to the capping nucleotide (i.e., in the initial nucleotide of the transcript) or in the first two nucleotides. The capping of the initial nucleotide has inspired a new term for molecular biology jargon: The site of transcription initiation by RNA polymerase II is called the **cap site.** As we have learned, this site is specified by the TATA box, some 30 nucleotides upstream.

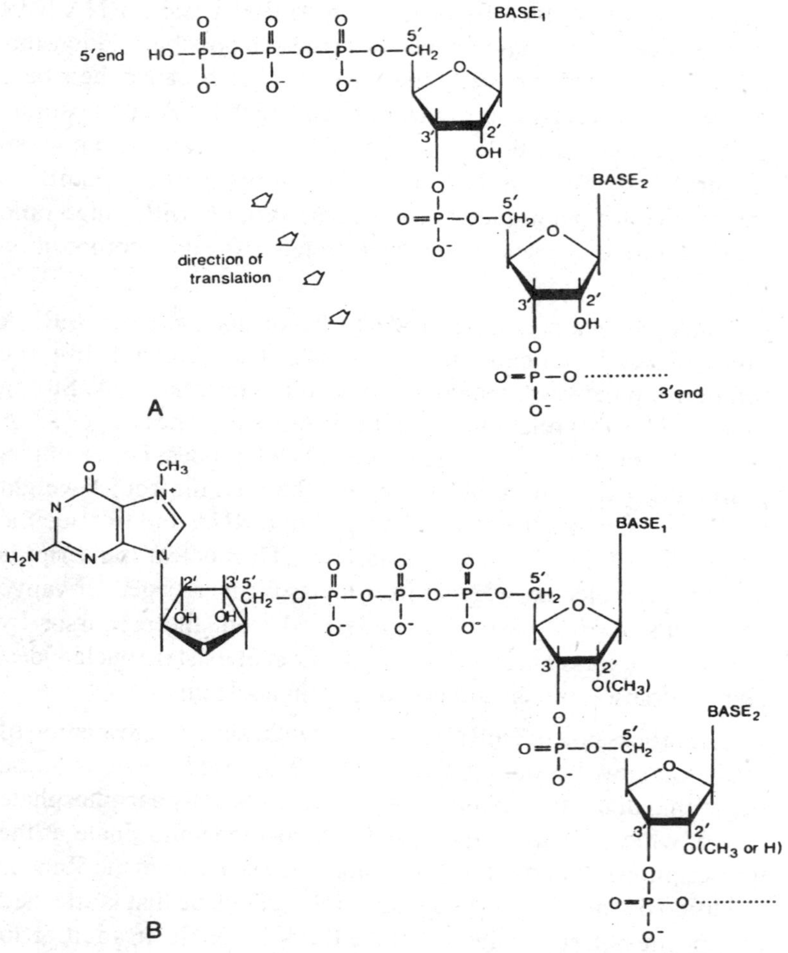

Figure 4.15 : The 5' terminus of mRNA.
A, Unmodified. B, Capped configuration.

The function of the 5' cap appears to be related directly to the role of the 5' end of the molecule in translation. Since translation is initiated at the 5' end of the messenger, the integrity of this portion of the molecule is crucial. A growing body of evidence suggests that 5' caps may be Vssential for the initiation process. An alternative function for the cap also has been proposed. Gedamu and Dixon (1978) suggest that the cap may protect messenger RNA from digestion by ribonuclease enzymes that degrade RNA from the 5' end. If degradation is rapid, initiation could not occur, and protein synthesis would be blocked.

The opposite, or 3' hydroxyl, end of a significant fraction of HnRNA and mRNA molecules has a series of about 150 to 250 adenylate residues. These regions are called **poly (A) tracts.** The poly (A) tract is added to the trailer sequence of the primary transcript, downstream from the site of the translation termination codon. Before addition of the poly (A), the trailer sequence is cleaved internally, and the poly (A) tract is added to the newly created 3' end of the transcript. A consensus hexanucleotide sequence AAUAAA is located upstream from the polyadenylation sites of transcripts (Proudfoot and Brownlee, 1976). This sequence is thought to be the signal for cleavage/polyadenylation. This interpretation is supported by the observations that (1) mutational transposition of the AAUAAA sequence results in a corresponding transposition of the polyadenylation site, which is 11 to 19 nucleotides downstream from the sequence; (2) mutational deletion of the AAUAAA sequence abolishes cleavage/polyadenylation; and (3) histone messengers, which lack poly (A) tracts, also lack the AAUAAA sequence.

The functional significance of poly (A) is not well understood. Its widespread occurrence in organisms of all phylogenetic levels in both the plant and animal kingdoms suggests that its presence on messenger molecules is important to their proper functioning. Most speculation concerning the role of poly (A) has centered on the possibility that it (1) functions in transport of mRNA from the nucleus to the cytoplasm, (2) helps to protect the messenger from degradation and hence extends its functional lifetime, (3) increases the efficiency of translation, or (4) functions in processing the HnRNA to mRNA. Defining a functional role for poly (A) has been complicated by the indisputable fact that some messengers lack poly (A) but are functionally normal. They are called poly

(A)$^-$ messengers, as distinguished from poly (A)$^+$ messengers. One welldefined poly (A)$^-$ messenger is histone mRNA (see preceding discussion).

Although a functional role for poly (A) in the cell has not been established, it has proved to be of inestimable value for investigators who need to prepare mRNA for experimental work. Because of the natural complementarity of adenine with uracil and thymine, the poly (A) associated with messengers will bind selectively either poly (U) or oligo (dT). In practice, affinity chromatography is used to isolate polyadenylated mRNA. A column of either poly (U) bound to Sepharose or oligo (dT) bound to cellulose is prepared. A solution of RNA is then passed through the column. Polyadenylated messengers bind, whereas other RPJAs pass through it. The messengers can then be eluted from the column by passing a reagent through it to break the bonds between poly (A) and poly (U) or oligo (dT). The messengers are collected in the column eluate.

One reason for the excess size of initial transcripts as compared to their mRNA derivatives is the presence of introns in the former, which are lacking in mature messengers. The mouse (3-globin gene transcript is an example. The nuclear transcript is about three times the size of the mature messenger, which is sufficient to encompass both the β-globin coding sequence and the two introns. This suggests that the precursor is a full-length gene transcript, whereas the messenger is obtained by removal of the introns from the precursor. Confirmation of this hypothesis has been obtained by ultrastructural comparisons of the structures formed by hybridizing the mouse β-globin gene with either the 15S precursor or lOS mRNA. The *entire* 15S (3globin RNA hybridizes with the β-globin gene. However, when the messenger RNA is hybridized with the gene, regions of the gene corresponding to the introns loop out because they are lacking in the messenger.

The removal of introns involves their excision, followed by splicing of the coding sequences. The excision and splicing steps must be performed with great precision to ensure the sequence integrity of the resulting messenger. The process is assumed to involve the bringing together of the 3' end of one coding sequence and the 5' end of the adjacent coding sequence. The intron would form a loop structure that could be excised enzymatically so that the adjoining coding regions could be ligated. Each intron begins with GU and ends with AG. These nucleotides are thought to provide the signal for precise excision/splicing.

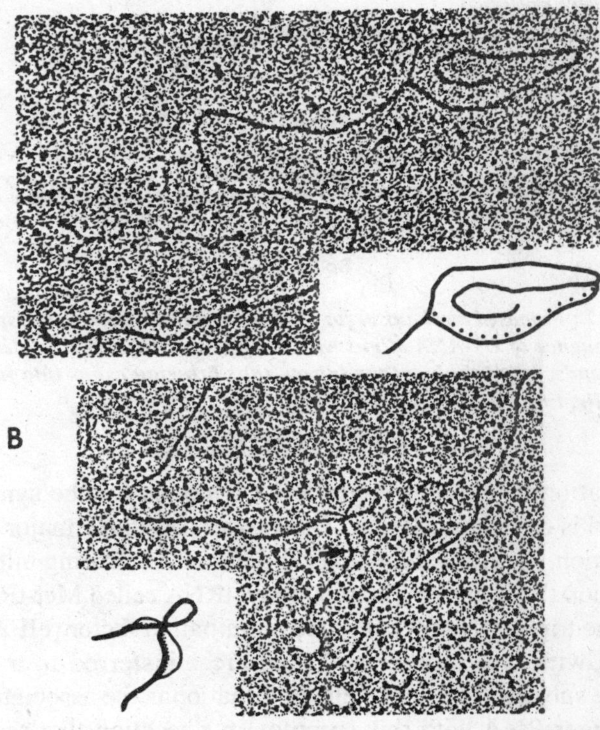

Figure 4.16 : The introns of the mouse β-globin gene. A, Electron micrograph of structure formed by incubation of partially denatured globin DNA with 15S transcript. Inset Interpretive drawing. X 120,000. (From S.M. Tilghman et al. 1978b. The intervening sequence of mouse β-globin gene is transcribed within the 15S B, Structure formed between globin DNA and globin mRNA. The large intron loops out because there is no complementary region in the mRNA. The arrow indicates a small loop corresponding to the small intron. Inset Interpretive drawing. The location of the small intron is not indicated on this drawing. Heavy lines represent double-stranded DNA; narrow lines represent singlestranded DNA; combined dotted lines-heavy lines indicate the DNA-RNA hybrids.

Excision/splicing may involve the participation of one of the small nuclear RNAs. This RNA, which is called U1-RNA has near its 5' end the base sequence ACCU. This portion of Ul-RNA could juxtapose by base pairing the splice junctions of HnRNA to facilitate cleavage and splicing.

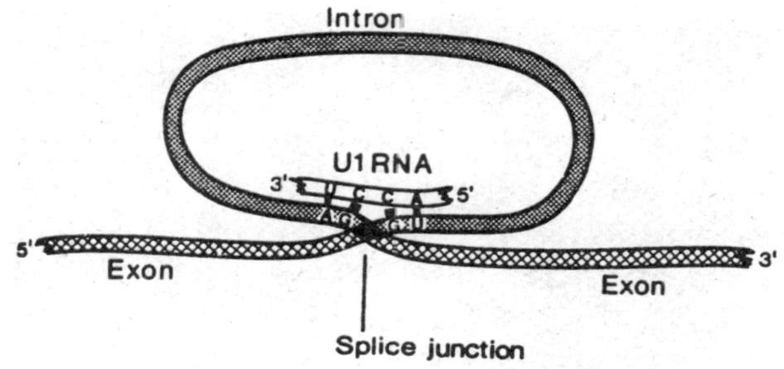

Figure 4.17 : Proposed mechanism for removal of introns from transcripts. The ACCU sequence of U1-RNA aligns with the GU and AG dinucleotides at the 5' and 3' ends, respectively, of the intron, thus bringing exons into juxtaposition for ligating.

PROTEIN SYNTHESIS

Utilization of the messenger RNA template for the synthesis of protein is called translation. This process has three major steps: (1) initiation, (2) elongation, and (3) termination. During initiation an initiation tRNA (a specific methionyl tRNA called Met-tRNA;), guanosine triphosphate (GTP), and an initiation factor (eIF2) first combine with one another and then are transferred to a small ribosome subunit. The next stage in initiation is the association of a messenger RNA with this complex in a reaction that requires ATP and initiation factors. Finally, the large ribosome subunit is added, completing the **initiation complex.** This reaction requires GTP and another initiation factor.

The Met-tRNA; recognizes an AUG codon near the 5' end of the messenger RNA. The initiation codon marks the start of the translatable portion of the mRNA molecule and specifies placement of a methionine amino acid at the beginning of the peptide. Each subsequent amino acid is specified as the codons are read in the 5' → 3' direction. Addition of these amino acids is elongation.

The transfer RNA bearing the next amino acid enters the ribosome, and a peptide bond is formed between this amino acid and the preceding one by the enzyme **peptidyl transferase.** The growing. peptide chain is then bound to the newly arrived tRNA, and the tRNA that was bound to the preceding amino acid is ejected from the ribosome. The peptide-tRNA complex is then translocated to the site formerly occupied by the ejected tRNA by movement of the mRNA. Translocation brings a new codon into the ribosome so that the next amino acid-tRNA complex can enter. The elongation process continues until a termination codon enters the ribosome. Termination is designated by either UAG, UAA, or UGA. The termination codon represents the 3' limit to the translatable portion of the mRNA. The entrance of the termination codon into the ribosome causes the completed polypeptide chain andd the ribosome to be released from the messenger.

Protein synthesis is generally accomplished by several ribosomes attached to a single mRNA. As translocation moves the initiation codon out of a ribosome, a new initiation complex can be formed. This process can be repeated several times; the number of ribosomes that can simultaneously read a messenger is roughly proportional to the length of the messenger. The complex of mRNA and associated ribosomes is a **polyribosome** or **polysome**.

As with transcription, inhibition of translation is an important experimental manipulation. Two commonly used inhibitors of eukaryotic translation are **puromycin** and **cycloheximide.** Puromycin is also frequently used in the preparation of messenger RNA from polysomes. This molecule enters the ribosome, causing premature termination of translation and release of ribosomes from the messenger. Thus, treatment of purified polysomes with puromycin allows for separation of the mRNA molecules from the ribosomes.

5

Gene Expression During Development

The evidence establishes that cells of an embryo, which are differentiating along distinct developmental pathways, may possess equivalent genetic information. This principle leads to the inevitable conclusion that the appearance of morphological and functional differences during development is due to differential utilization of genetic information. However, as with any generalization concerning the living world, there are exceptions to the rule of genome constancy during development. Having outlined in the previous chapter the modes of information utilization, we shall now describe how the utilization of different portions of the genome can be modulated to produce divergent cell types.

As we pointed out earlier, the uniqueness of any cell is due to the functional proteins it contains. Obviously, every cell requires certain essential enzymes and structural proteins to carry on the "housekeeping" functions of normal metabolism. But the *raison d'etre* of any type of cell in a multicellular organism is to carry out some special function that is not shared by any other cell type, and every specialized function requires a special set of enzymes and structural proteins. Cell specialization therefore requires a developmental commitment to restrict the protein synthetic machinery to the production of a special protein "set." Consequently, cell differentiation is marked by dynamic changes in the messenger RNA population in the cytoplasm.

The ultimate change in protein synthetic patterns is speciali-zation' for the synthesis of one major functional protein. This kind of restriction occurs when the messenger population becomes dominated by a single messenger species. For example, differentiation of an erythroid cell is accompanied by the preferential accumulation of globin messenger. A corollary to the presence of specific messengers in specialized cells is that these messengers are lacking in other specialized cell types; for example, globin messenger is unique to erythroid cells, which lack the messengers that characterize other cell types. The appearance and accumu-lation of specialized messengers provide a useful means of moni-toring cell differentiation.

The translation of unique messengers by specialized cells is not the cause of differentiation, but its result. There is no single mechanism that causes these restrictions-regulation of gene expression could be exerted at each step in cellular information processing. In this chapter we shall examine these regulatory mechanisms. The most decisive step in information processing is transcription. Obviously, if a gene were not transcribed, there would be no need for posttranscriptional regulation. What mechanisms determine which genes are to be transcribed into RNA? In fact, is there any discrimination at all at this level, or are all genes transcribed, only to be thwarted at a later step? Since cytoplasmic RNA is derived from a more complex population of nuclear RNA, some discrimination process must determine which RNA can enter the cytoplasm ' How are specific RNA sequences selected from the precursor population and thus saved from degradation by the nuclease enzymes? Finally, does a cell translate all of its messenger RNA, or are there mechanisms that channel specific ones onto ribosomes?

The answers to these questions are not all known, but enough is known to make an exciting story. Although much of this information has been gathered by molecular biologists, the results of their investigations are vital to the developmental biologist's understanding of cell differentiation. It is important to keep in mind that what is significant to developmentalists is the process of *change.* Development is a dynamic process, and comparisons of static differences are relevant to development only insofar as they relate to how different cell types are produced and how progressive stages of differentiation are achieved.

EVIDENCE FOR DIFFERENTIAL GENE TRANSCRIPTION

The majority of the nucleotide sequences in the genome are inactive in transcription in most cells. Selection of the "active" versus the "inactive" genes is considered by many investigators to be a critical step in controlling cell differentiation. However, experimental demonstration of differential transcription is difficult. The evidence for differential transcription has been derived from cytological studies and, more recently, from the fast-moving field of molecular biology.

Polytene Chromosomes

One of the most dramatic demonstrations of differential transcr-iption is the morphological change occurring in certain loci of dipteran polytene chromosomes when they become functionally active. Polytene chromosomes are found in certain enlarged cells of the larvae of some dipterans, including *Drosophila* and *Chironomus*. Dipterans are somewhat unusual, since homologous chromosomes pair and remain perfectly aligned in somatic cells during interphase. In larval salivary glands, midgut, and malpighian tubules of *Drosophila* and *Chironomus*, repeated DNA replication of both homologues follows somatic pairing. The numerous replicates (chromatids) do not separate but remain attached side by side, forming one extremely large structure-a polytene chromosome. Polytene chromosomes can attain a cross-sectional width as much as 10,000 times that of a normal interphase chromosome. The increased size and the parallel alignment of the chromatids reveal structural details that cannot be seen in normal interphase chromatin.

The polytene chromosomes of the *Drosophila* larval salivary glands. The four chromosomes radiate out from a structure called the **chromocenter,** which results from an aggregation of the centromeres of all the chromosomes. The chromosomes have a striated appearance of bands alternating with interband regions. The bands are visible because the concentration of DNA is greater than in the interbands, due to local folding of each chromatid. Each of these folded domains is called a **chromomere.** The parallel alignment of chromomeres on the individual strands produces the dense bands.

Beermann (1952) observed that some bands may exhibit a swollen or **puffed** appearance under certain conditions. A puff results from an unfolding of the chromomeres that constitute a band. Puffing may be slight, with the only apparent change being a small enlargement and diffusion of the band, or it may be extreme, with the uncoiling of the chromomeres to form large loops. The larger puffs found in *Chiron*omus are called **Balbiani rings.** Beermann proposed that the puffs represent sites of intense transcriptional activity. This hypothesis is supported by several lines of evidence. For example, examination of Chironomus salivary gland chromosomes, using the Miller technique for observing transcription by electron microscopy, reveals regions that are assumed to be Balbiani rings that show chromatin loops bearing highly active transcription units. These loops are presumably formed by the unfolding of individual chromomeres. Also, puffs readily incorporate radioactive RNA precursors, as demonstrated by au-

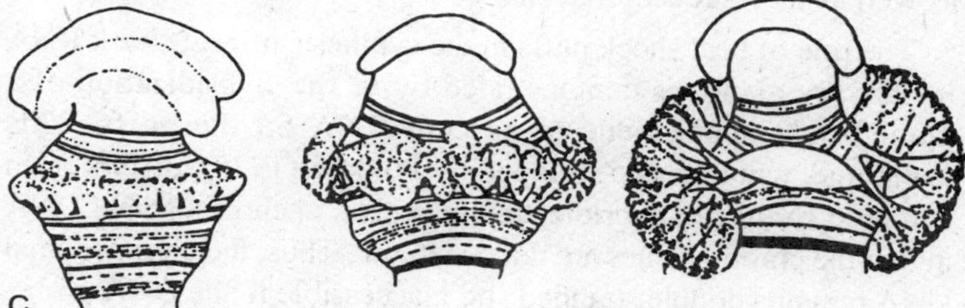

***Figure 5.1 : Diagrammatic representation of one of the three large puffs (Balbiani rings) of chromosome 4 characteristic of all salivary gland nuclei of* Chironomus tentans *and* Chironomus pallidivittatus.**

toradiography. Furthermore, one can correlate the amount of incorporation of label with the extent of puffing; a puff site that has undergone experimentally induced puff regression demonstrates a commensurate reduction in ^{3}H-uridine incorporation. Regression of puffing can be induced with a variety of agents. This is illustrated by incubation of *Chironomus* salivary glands with the RNA polymerase II and III inhibitor, α-amanitin, which causes regression of Balbiani rings. Incorporation of ^{3}H-uridine in the regions of puff regression (as with all nonnucleolar chromosome regions) has been eliminated. These results suggest that these regions are transcriptionally active when they are decondensed (i.e., puffed) and suppressed when they are condensed.

Confirmation of this interpretation is provided by experiments in which puffing is induced, and this morphological change is correlated with the synthesis of specific transcripts. An experimental procedure that facilitates this kind of analysis is treatment of *Drosophila* with high tem eratures. Briefly shifting the organisms from normal temperatures (approximately 25°C) to 37°C elicits the formation of a discrete set of puffs and the synthesis of a specialized set of mRNAs and proteins. These proteins are called heat shock proteins. At the same time, the expression of other gene loci is inhibited; all preexisting puffs regress, most nonheat shock RNAs are no longer produced, and the translation of preexisting messengers is halted. The same effects on RNA and protein synthesis are also observed in other *Drosophila* tissues and cells as well as in tissue culture cells.

The role of heat shock puffs in the synthesis of heat shock RNA in salivary glands is demonstrated by *in situ* hybridization of a cloned heat shock gene to nascent RNA on the puffs. This technique, which was developed by Pukkila (1975), differs from that used for *in situ* hybridization of probes to chromosomal DNA in that the chromosomes are not denatured. Thus, the chromosomal DNA remains double stranded and inaccessible to the DNA probe. The hybridization detected by autoradiography is therefore due to annealing between the probe DNA and the nascent RNA. Two of the heat shockinduced puffs contain RNA that hybridizes with the cloned probe. No hybridization is detected without heat shock. Consequently, heat shock induces transcription at these loci.

Comparisons of puffing patterns of polytene chromosomes

during normal development illustrate that transcription is subject to both spatial and temporal regulation. Examination of puffs in different cell types reveals that each type has a unique puffing pattern. For example, Balbiani ring 2 is formed in salivary gland chromosomes of *Chironomus* tentans but not in other tissues having polytene chromosomes, such as the matpighian tubules. Temporal regulation is illustrated by examination of puffs of *Drosophila* in the same tissue at different stages of development; the pattern of transcription changes as development proceeds. Developmental regulation of puffing is due to the **moltiVg** hormone **ecdysone,** which causes puffing of some gene loci and regression of puffs at other sites. We shall discuss this function of ecdysone.

Facultative Heterochromatin

Visual evidence for differential transcription is also provided by facultative heterochromatin, in which one chromosome of a pair becomes heterochromatic during development, and hence, only one gene of each allele pair is active in transcription. In addition to its genetic inactivity, facultative heterochromatin has other properties in common with constitutive heterochromatin; that is, both types of heterochromatin remain condensed during interphase and are late replicating. Their fundamental difference is that constitutive heterochromatin is composed of highly repetitive DNA, whereas facultative heterochromatin is not. Instead, the latter type behaves as heterochromatin because of a still undefined developmental event that renders it heterochromatic.

A classical example of facultative heterochromatin involves one of the X chromosomes of female mammals. Although females contain two X chromosomes (and hence two copies of each X-linked gene) and males carry only one, both sexes produce similar amounts of proteins that are coded by genes located on the X chromosome. This is achieved by a mechanism called **dosage compensation.** Mary Lyon (1961) proposed that dosage compensation occurs by random heterochromatization of either the paternal or maternal X chromosome during development (the Lyon **hypothesis).** After the initial inactivation occurs in an embryonic cell, the same X chromosome remains inactive in that cell and all its mitotic derivatives during the lifetime of the organism. The heterochromatic X can be seen as a darkly staining structure called the Barr body lying near the nuclear envelope in interphase nuclei.

Since heterochromatization is random, females that are heterozygous for genes on the X chromosome have a **mosaic** phenotype; that is, somecells will have the paternal chromosome inactivated, whereas others will lose the function of the maternal chromosome. For example, the gene for glucose-6-phosphate dehydrogenase (G-6-PD) is located on the X chromosome in humans. Two different alleles of the G-6-PD gene have been found in the American black population. These alleles produce enzymes with different electrophoretic mobilities. The alleles are designated A and B; A produces an enzyme that migrates rapidly during electrophoresis, whereas B produces a slowly migrating enzyme.

Individual cells of a female heterozygous for the two electrophoretic forms produce either the A form or the B form of the enzyme but not both. She is, therefore, a phenotypic mosaic, with small patches of cells that produce either form of the enzyme scattered throughout her body.

This can be demonstrated by sampling skin cells, which can be conveniently removed and placed in culture. After the cells are established and have multiplied *in vitro,* single cells are selected to form new cultures, which are clones of genetically identical cells. Cells from the clones are then homogenized, and the G-6-PD enzymes are resolved by electrophoresis. Each clone has only the A or the B form of the enzyme, never both.

Molecular Evidence

Differential transcription is dramatically demonstrated by polytene chromosome puffing patterns and by formation of facultative heterochromatin. Most chromatin, however, is not so obliging as to provide such obvious clues of transcriptional activity. Therefore, we require molecular techniques to reveal modulation of transcription so that we may examine transcriptional regulation in detail. Differential transcription is indicated by the presence of different transcripts in different cell types. We know, for example, that specialized cells preferentially accumulate certain transcripts.

Molecular probes, such as cloned cDNA, have the specificity and sensitivity to detect low levels of transcripts, and thus can distinguish between their presence and absence. Hybridization analyses between probes and RNA extracted from cells reveal that (1) the transcripts for cell-specific proteins increase in amount during cell differentiation and (2) cells specialized for other functions do not have significant numbers of these same transcripts. For example, globin transcripts are present in chick embryo erythropoietic tissue but are not found in significant amounts in either the nuclear or cytoplasmic RNA of nonerythropoietic tissue. The results of a large number of experiments with probes for transcripts encoding cell-specific proteins strongly suggest that genes for such proteins are regulated at the transcriptional level.

TRANSCRIPTIONAL CONTROL MECHANISMS

The preceding evidence indicates that gene expression can be regulated at the transcriptional level. We shall now examine how transcriptional regulation could occur. This discussion will center on chromatin itself and on extrinsic factors that affect cell differentiation.

Properties of Transcribed Chromatin

The discovery of the basic structure of chromatin has made it possible to attempt to correlate chromatin structure with its function in transcription. Specifically, does the configuration of chromatin differ between transcribed and nontranscribed chromatin? If so, what is the nature of the difference, and how is it regulated? The logistical problems,) of transcription suggest that modifications to chromatin would facilitate RNA synthesis. For example, the RNA polymerase molecule is itself about the same size as a nucleosome. The DNA is wound around the outside of the nucleosome core particle. The movement of the large polymerase molecule along the DNA would probably be facilitated if some modification of this configuration were effected. Furthermore, the two strands of the DNA helix are separated in the region that is transcribed by the polymerase. Strand separation may be facilitated by a modified nucleosomal configuration. Alternatively, strand separation itself would most likely alter nucleosomes.

The most useful tools for distinguishing between transcribed and nontranscribed chromatin have been nuclease enzymes such as micrococcal nuclease and DNase I. The use of these enzymes is illustrated by studies on chick globin and ovalbumin genes. These genes are relatively resistant to digestion by micrococcal nuclease in all tissues, demonstrating that they are packaged into nucleosomelike particles. However, brief treatment of erythrocyte nuclei with DNase I degrades much of the DNA that hybridizes with globin cDNA (i.e., the globin genes). As a control, ovalbumin cDNA was used to probe the presence of ovalbumin genes in nuclease-treated erythrocyte and fibroblast nuclei. These genes are inactive in both cell types, and as shown in the graph, they are not digested by the enzyme. The globin gene is also inactive in fibroblasts. In a result that is consistent with the ovalbumin data, DNase I was found to have no effect on globin genes of fibroblast nuclei. Differential sensitivity to DNase I reflects not only the transcriptional activity of genes but also latent transcriptional competence; that is, globin genes of mature erythrocytes that are no longer active in transcription retain this property.

Garel and Axel (1976) investigated the sensitivity of ovalbumin genes to DNase I in hen oviduct nuclei and liver nuclei; the ovalbumin genes are transcribed in the former tissue but not in the latter. In agreement with the previously described studies, they found that brief treatment with DNase I preferentially digests the ovalbumin genes in nuclei in which they are transcriptionally active (i.e., in oviduct nuclei, but not in liver nuclei). The preferential sensitivity of chromatin composed of transcribed genes suggests that the nucleosomes of this chromatin are in a modified configuration that renders the DNA more susceptible to nuclease digestion. Perhaps this configuration facilitates transcription by making the DNA accessible to RNA polymerase, but this remains to be verified.

The demonstration of a modified configuration for transcriptionally active chromatin has stimulated a search for the structural basis of this difference. It is hoped that this search will yield important clues regarding the regulation of gene expression. Attention has been focused on the nonhistone chromosomal

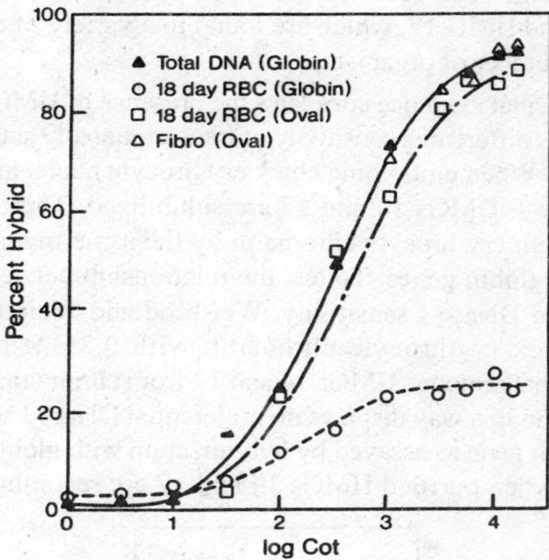

Figure 5.2 : Preferential digestion of active genes by pancreatic deoxyribonuclease I. Red blood cells (RBC) were obtained from 18-day chick embryos, and fibroblasts (fibro) were dissected from 11-day chick embryos and grown in culture. Nuclei were isolated from these cells and digested with DNase I until 10 to 20% of the DNA was soluble in trichloroacetic acid. The remaining DNA was extracted for hybridization studies. Total DNA was prepared directly from 18-day red cell nuclei. DNA samples were hybridized with either ovalbumin cDNA or globin cDNA. The percentage of hybridized cDNA is plotted as a function of the Cot. At the completion of the reaction between total DNA and globin cDNA, the curve plateaus since all of the cDNA has reacted with complementary sequences. However, after treatment of red cell nuclei with the nuclease enzyme, the hybridization reaction plateaus at a lower level. The lower plateau indicates that fewer globin sequences are available to react with the cDNA. The difference between plateau levels is a measure of the globin sequences digested by the enzyme. By comparison, the plateau levels with ovalbumin cDNA are not reduced in either red cell or fibroblast DNA from nuclease-treated nuclei. Therefore, the ovalbumin sequences are not digested by the enzyme.

proteins, specifically a fraction known as the **high mobility group (HMG).** These small proteins have high mobility on polyacrylamide gels-hence their name. They have a high percentage of both acidic and basic amino acids, making them highly charged molecules. They are easily extracted from chromatin with 0.35 M NaCl. Two

HMG proteins have received considerable attention. These are HMG 14 and HMG 17, which are found in a variety of cell types in a large number of organisms.

Experimental evidence correlates the presence of HMGs 14 and 17 with the , differential sensitivity of transcriptionally active genes to DNase I. When embryonic chick erythrocyte nuclei are treated with DNase I, HMGs 14 and 17 are solubilized. Thus, they are released from erythrocyte chromatin by the same treatment that solubilizes globin genes. To test the relationship between HMG proteins and DNase I sensitivity, Weisbrod and Weintraub pre-treated chick erythrocyte chromatin with 0.35 *M* NaCl-the procedure that removes HMGs 14 and 17 from chromatin. Chromatin treated in this way displays no preferential DNase I sensitivity of the globin gene as assayed by hybridization with globin cDNA. However, when purified HMGs 14 and 17 are recombined with

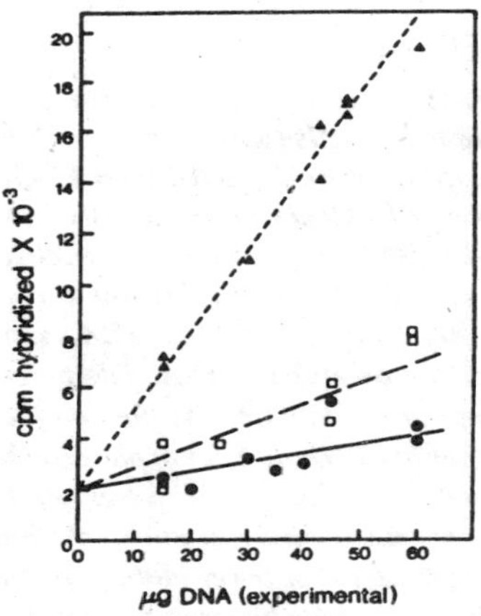

Figure 5.3 : Reconstitution of DNase I sensitivity of chick embryo erythrocyte chromatin with HMG proteins. The chromatin was treated with 0.35M NaCl to remove HMGs 14 and 17 and then was separated into separate aliquots. One portion was treated with DNase I (line with solid triangles), and the other was reconstituted with HMGs 14 and 17 before DNase I treatment (line with open squares). As controls, nuclei were treated with DNase I to 10% trichloroacetic acid solubility (line with closed circles). Each sample was then hybridized with excess globin cDNA (total globin cDNA was 100,000 cpm) to measure relative numbers of globin genes in the chromatin. There is a clear correlation between the presence of HMGs 14 and 17 and DNase I sensitivity.

HMG-depleted chromatin, DNase I sensitivity is restored. This result indicates that the globin genes in HMG-depleted chromatin retain some property that allows them to bind HMG proteins selectively, which renders them DNase I sensitive. Furthermore, HMGs 14 and 17 from brain nuclei will restore DNase I sensitivity to HMG-depleted erythrocyte chromatin, but erythrocyte HMGs will not render the globin genes of brain chromatin sensitive to DNase I. Thus, the HMG proteins are not responsible for the selectivity; this property is retained by some other component of the chromatin. In a more recent study, Weisbrod et al. (1980) demonstrated that actively transcribed genes in *general* are rendered sensitive to DNase I by HMG proteins-an indication that this property is not unique to globin genes.

The experiments we have just discussed suggest that differential nuclease sensitivity is dependent upon the association of HMGs 14 and 17 with active genes. However, this association itself must be a consequence of some more basic property of transcribed genes that causes HMGs 14 and 17 to interact with them. Two possibilities to be considered are modifications to histones and to DNA itself. Perhaps significantly, the known modifications to histones occur in regions of the molecules that interact with DNA. Thus, changes in histones could reduce the compaction of DNA within nucleosomes and facilitate transcription. One histone modification that is known to be correlated with transcriptional capacity of chromatin is acetylation of the lysine residues of the core histones. Chromatin containing a high level of histone acetylation is very sensitive to DNase I, which we have learned is a property of transcriptionally active chromatin. It has been proposed that HMGs 14 and 17, which are also associated with transcriptionally active chromatin, may help to maintain the high level of acetylation in this chromatin.

Recently, a great deal of interest has been generated concerning the possibility that modifications to DNA bases may be correlated with transcriptional activity of genes. The most common modification is postreplication methylation of cytosine to produce 5-methylcytosine (m^5C). Methylation occurs preferentially in the dinucleotide sequence m^5CG and is detected by a comparison of the cleavage patterns of the restriction enzymes *Hpa II* and *Msp I*. Both these enzymes recognize sites containing *CCGG* sequences. However, *Hpa* 11 cannot cleave Cm^S*CGG*, whereas *Msp* I cleaves both *CCGG* and Cm^S*CGG*. Thus, *Msp* I is used to identify CCGG sites in genes, and methylation of these sites is detected by comparison with the digestion pattern obtained with *Hpa* II.

In a typical experiment, total DNA is restricted with these enzymes, the DNA fragments are resolved by agarose gel electrophoresis and transferred by Southern blotting to nitrocellulose. The DNA fragments containing genes of interest are then visualized by hybridization to a labeled probe. The figure compares the restriction patterns obtained for the α-fetoprotein (α-FP) gene from yolk sac endoderm and mesoderm. The α-FP gene is expressed in endoderm but not in the mesoderm. Msp I cleaves the α-FP genes from both tissues into five fragments (labeled a-e). The same five fragments are generated with *Hpa II* treatment of endoderm DNA. However, treatment of mesoderm DNA with *Hpa II* releases only one fragment. These contrasting results demonstrate that the CCGG sites are methylated in mesoderm but hypomethylated (i.e., the level of methylation is reduced) in endoderm-the tissue in which the gene is expressed.

Results with a number of genes indicate that in cells expressing a specific gene, CCGG sites are hypomethylated, whereas the highest levels of methylation are found in tissues in which the gene is not expressed. These observations support the hypothesis that demethylation may be necessary for gene expression during development.

The demethylation hypothesis is also supported by the results of experiments with the nucleoside analogue 5-azacytidine. This drug, which causes demethylation of cytosine residues in DNA, has been shown to promote the expression of previously unexpressed genes. An other dramatic example of the effects of 5-azacytidine is the activation of genes on heterochromatic mammalian X chromosomes in somatic cell hybrids with this drug.

One of the most intriguing consequences of the demethylation hypothesis for regulation of gene expression is that it satisfies the important requirement for stable transmission of the pattern of gene expression of a cell to its progeny. Methylation of the appropriate cytosine residues occurs enzymatically after replication of DNA, conserving the methylation pattern of the parental strands in each daughter strand. Another important property of methylated CG sequences; that is, the methylation pattern is symmetrical on the two strands of the DNA double helix. The symmetrical methylation makes possible the clonal inheritance of the exact methylation pattern at cell division, presumably by means of a methylase enzyme that recognizes the half-methylated sites that exist after replication and restores the fully methylated symmetrical pattern.

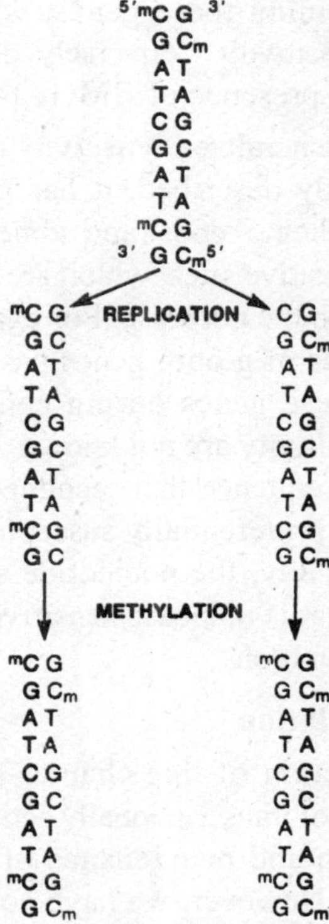

Figure 5.4 : Restoration of the parental methylation pattern following DNA replication. Note that of the three CG dinucleotides only those methylated before replication show methylation in their complementary strands after replication.

We shall now summarize the demethylation hypothesis. Genes may be retained in the unexpressed state by maintenance of the methylation of certain cytosine residues within the genes or in their vicinity.

When cells with methylated cytosines divide, the methylation pattern is inherited by their progeny. The cell determination events that lead to differentiation of specific cell types would cause demethylation of these critical cytosine residues in particular genes, allowing these genes to be expressed. When determined cells divide, they retain the pattern of methylation-demethylation that maintains the differentiation state in their progeny. Perhaps as a consequence of demethylation, HMGs 14 and 17 associate with the chromatin containing these genes, which in turn displays enhanced DNase I sensitivity. Conversely, demethylation might be a consequence of the presence of HMGs 14 and 17.

In addition to the generalized sensitivity of transcribed genes to DNase I as previously described, it has been shown that such genes are flanked at their 5' ends (and sometimes at their 3' ends) by so-called hypersensitive sites, which are preferentially cleaved by light digestion with the nuclease. For example, the appearance of hypersensitive sites of globin genes has been correlated with the expression of these genes during cell differentiation. The reasons for hypersensitivity are not known. One possibility is that a special nucleotide sequence that recognizes and interacts with regulatory proteins is preferentially susceptible when the proteins are present. Alternatively, the nucleotide sequence may have a peculiar base order that is nuclease sensitive when the chromatin is in the active configuration.

Specific Gene Regulation

The preceding section of this chapter outlines some of the generalized properties of transcriptionally active chromatin that may facilitate transcription and maintenance of the transcriptionally active state of genes. However, we have not addressed the problem of the regulation of *differential* gene expression. How are specific genes selected to be transcribed from the vast library of available sequences? It is this selectivity that allows for the differentiation of cells with specialized properties. It is generally assumed that specific nonhistone chromosomal proteins interact with gene regulatory sequences to facilitate gene expression. However, proof of this hypothesis is a very difficult proposition. One problem is that many, if not most, of the NHC proteins show little tissue specificity (e.g., HMGs 14 and 17). Moreover, since most structural genes are present as single copies, putative regulatory proteins may be present in amounts too small to be detected by current analytical techniques. The 5S RNA genes of *Xenopus laevis* are highly reiterated and for this reason (among others) have provided a favorable system to search for a specific regulatory protein.

Transcription of 5S RNA genes is regulated by a protein called TF IIIA. The function of TF IIIA is demonstrated by *in vitro* transcription analyses. Cloned 5S RNA genes are transcribed *in vitro* only in the presence of TF IIIA. When TF IIIA is combined with the 5S RNA genes it binds to the gene promoters to facilitate transcription. The specificity of TF IIIA is demonstrated by the fact that it has no effect on transcription of tRNA genes, which-like the 5S RNA genes-are transcribed by RNA polymerase III. The presence of TF IIIA is thought to be necessary for the transcription of 5S RNA genes during oogenesis. TF IIIA is reduced to very low levels in eggs after ovulation; this deficiency is apparently responsible for the absence of 5S RNA synthesis following ovulation and for preventing the activation of the zygote's oocyte-type 5S RNA genes during early embryonic development. Thus, regulation of the cellular level of TF IIIA may indirectly regulate expression of the oocyte-type 5S RNA genes. Additional work with 5S RNA genes and the discovery of other gene-specific regulatory proteins and their effects on chromatin will surely yield additional valuable clues about regulation of transcription.

The distinct properties of transcriptionally active chromatin we have described must be taken into consideration in any experiments attempting to modify previously established patterns of gene expression, such as in nuclear transplantation and somatic cell hybridization studies. The failure of a nucleus to show an alternate pattern of gene expression does not necessarily mean that the DNA nucleotide sequence has been changed. It may simply mean that the pattern of expression imprinted on the chromatin (by the presence of NHC proteins, acetylation of histones, or methylation of cytosine) is stable and cannot be reversed under the particular experimental conditions. A more complete understanding of the regulation of gene expression will be forthcoming when we discover more not only about the establishment of patterns of gene expression but about reversal of that process as well.

Coordination of Transcription

Although immediate control over transcription may be mediated by chromosomal proteins, it is still necessary to account for the chronological and spatial patterns of gene expression that occur during orderly development. The transcription of a gene or set of genes may be mediated by the presence of a factor produced by yet another gene in the same cell. For example, the transcription of 5S RNA genes is dependent upon the presence of TF IIIA, which is itself a product of gene expression. The transcription of the oocyte-type 5S RNA genes during oogenesis and after fertilization appears to be regulated by the cellular level of TF IIIA.

The functional properties of cells in embryos may be influenced by specialized cytoplasm that is heterogeneously distributed in eggs **(nucleocytoplasmic interactions)** and by interactions between adjacent cells **(cell-cell interactions).** These effects could be exerted at any level, and demonstrating regulation at the level of transcription is not an easy task. One experimental system that is amenable to such an analysis is somatic cell hybridization. In one such experiment, Deisseroth and Hendrick (1999) have shown that human a-globin genes from nonerythroid cells are transcribed when fused with mouse erythroid cells. Thus, some cytoplasmic factor or process in the mouse erythroid cells must have the capacity to activate quiescent a-globin genes.

Another significant influence over cell function during development is the endocrine system. Hormones have a dramatic effect on organisms and the state of differentiation of their cells. Experimental analysis of hormonal effects is facilitated by their ease of administration, both *in vivo* and *in vitro*. Consequently, transcriptional regulation by hormones has been studied extensively.

We shall analyze the results of some of these experiments subsequently.

REGULATION OF PUFFING IN POLYTENE CHROMOSOMES BY ECDYSONE

Direct hormonal control of transcription is illustrated by the modulation of the puffing activity of polytene chromosomes in dipteran larvae. Detailed examination of the polytene chromosomes has revealed the existence of specific patterns of puffing activity that vary with cell type and stage of development. The larval phase of the dipteran life cycle is characterized by a series of stages, each terminated by a **molt.** The final molt (sometimes called metamorphosis) results in formation of the immobile pupa. A complex pattern of regression of some puffs and appearance of new puffs occurs just preceding larval molts or metamorphosis.

These abrupt changes in puffing pattern correspond to the time of release of the hormone ecdysone, which triggers molting. This correlation led Becker (1989) to propose that ecdysone is directly responsible for regulation of puffing activity. Since ecdysone is released from the prothoracic gland, which is located in the first thoracic segment, Becker (1992) was able to ligature *a Drosophila* larva prior to hormone release so that the anterior portion of the elongate salivary gland would be exposed to the hormone, but the posterior portion would be isolated from the source of the hormone. When Becker examined the polytene chromosomes, he found that the chromosomes in the anterior part of the gland showed a normal stage-specific puffing pattern after ecdysone release, whereas those in the posterior portion retained the pattern existing before ligation. In a further series of experiments Becker transplanted salivary glands between animals of different stages and found that puffs characteristic of the host animals were induced. These experiments confirmed Becker's proposal that stage-specific puffs are regulated by ecdysone.

The role of ecdysone has been demonstrated more directly by experiments in which the hormone was injected into larvae, resulting in premature appearance of stage-specific puffs. More recently, isolated *Drosophila melanogaster* salivary glands have been exposed to ecdysone in *vitro*. The chromosomes of hormone-treated glands apparently progress through a normal puffing pattern. A chromosomal region from salivary glands treated with ecdysone *in vitro*. The region corresponds to one whose normal in *vivo* puffing pattern. A comparison of these two figures shows that hormone treatment results in a pattern that is virtually identical to that found in *vivo*. Note in particular the sequence of puffing-regression-puffing shown by bands 74EF and 75B.

Investigators have recently determined that certain ecdysone-responsive genes code for so-called glue polypeptides that are secreted from the salivary glands to attach the pupal case to the substrate. Furthermore, some of these genes have now been cloned, providing probes for transcripts of these genes and a means for detailed analyses of the genes. Therefore, this should prove to be an excellent experimental system for analyzing regulation of transcription at the molecular level. Recent evidence suggests that ecdysone binds specifically to ecdysone-responsive chromosomal regions. However, the subsequent events in transcriptional regulation are unknown. The cloning of ecdysone-responsive genes may be the technological breakthrough that will facilitate this analysis.

REGULATION OF TRANSCRIPTION BY VERTEBRATE STEROID HORMONES

Another important system for analyzing hormonal regulation of transcription is the steroid hormone response of vertebrates. Steroids evoke changes in patterns of protein synthesis. These changes are apparently mediated, at least in part, by modifications in transcriptional patterns. This conclusion was originally based upon the fact that steroid responses are inhibited by a-amanitin and actinomycin D and is strongly supported by more direct evidence, which we shall discuss later.

One of the most popular experimental systems for studying the effects of hormones on transcription is the chick oviduct response to estrogen and progesterone. Estrogen is required for the differentiation of the tubular gland cells of the oviduct, the cells that produce the eggwhite proteins-ovalbumin, conalbumin, lysozyme, and ovomucoid.

The major egg-white protein is ovalbumin, which makes up as much as 50% of the protein that is synthesized by the gland cells. Estrogen is not present in measurable amounts in immature chicks. Hence, the effects of the hormone on gland cell differentiation and ovalbumin synthesis can be studied by hormonal injection. The oviducts of immature chicks can be induced to develop by daily administration of estrogen, which causes proliferation and differentiation of the tubular gland cells.

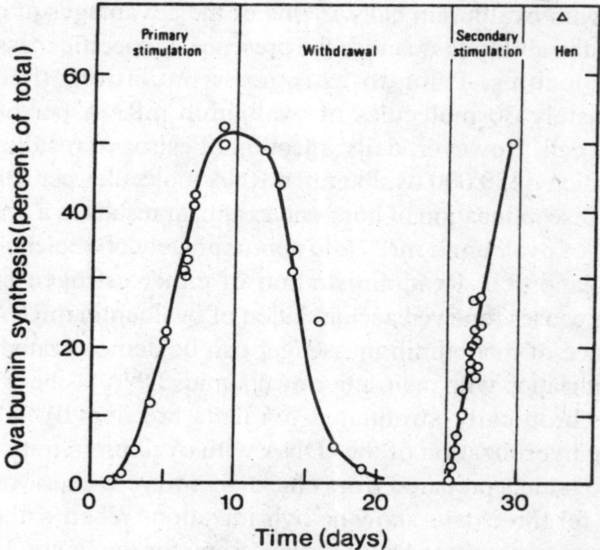

Figure 5.5 : Effect of estrogen on the relative rate of synthesis of ovalbumin in chick oviduct during primary stimulation, withdrawal, and secondary stimulation. Immature 4-day old chicks were injected with estrogen daily (primary stimulation), and after 10 days without estrogen administration (withdrawal), administration was resumed (secondary stimulation). Results are presented as per cent of total protein synthesis.

Ovalbumin synthesis is first detectable in the oviduct about 18 hours after hormone administration and reaches a plateau after about ten days. If hormone treatment is interrupted, ovalbumin synthesis declines to undetectable levels, although some of the tubular gland cells remain. Resumption of estrogen administration or administration of progesterone **(secondary stimulation)** reinitiates ovalbumin synthesis in the preexisting gland cells after about four hours. Secondary stimulation provides some experimental advantages over primary stimulation for studying hormonal effects: The oviduct is much larger than it is during primary stimulation, and the initial events in ovalbumin synthesis are unaffected by cell division, unlike in primary stimulation.

The increased ovalbumin synthesis promoted by steroids is the direct result of an accumulation of ovalbumin messenger RNA in the cells of hormone-treated tissues. This is shown by hybridization analysis with ovalbumin cDNA. One of the advantages of cDNA studies is the ability to quantify the presence of specific messenger RNA molecules. Prior to estrogen stimulation, there are approximately 30 molecules of ovalbumin mRNA per oviduct epithelial cell. However, daily injection of estrogen results in the accumulation of 50,000 ovalbumin mRNA molecules per cell after 18 days. Discontinuation of hormone treatment results in a dramatic reduction of ovalbumin mRNA to approximately 60 molecules per tubular gland cell. Readministration of either estrogen or progesterone causes renewed accumulation of ovalbumin mRNA. The pres- ; ence of ovalbumin messenger can be demonstrated by in situ hybridization with radioactive ovalbumin cDNA. Tubular gland cells of chronically stimulated oviducts are heavily labeled, reflecting hybridization of the cDNA with ovalbumin transcripts. Cells of oviducts prepared from chicks that were withdrawn from estrogen for three days show no hybridization. When withdrawn chicks were restimulated with progesterone for ten hours, labeled cells were detected once again.

Do the differences in ovalbumin messenger accumulation reflect changes in the rates of ovalbumin mRNA synthesis in these different physiological states, or could the differences have other explanations such as different turnover rates for the transcripts, with transcripts in cells from hormone-withdrawn chicks being degraded rapidly following their synthesis? The synthetic rates have been determined by hybridization of radioactively labeled messenger RNA to cloned ovalbumin cDNA. Chronic injection of estrogen leads to high levels of ovalbumin messenger synthesis that remain constant as long as hormone administration is continued. The synthetic rate drops rapidly during estrogen withdrawal to become undetectable by 60 hours. Readministration of hormone causes a rapid return to the maximal level of synthesis. Thus, the steroids exert their effects on the level of ovalbumin gene transcription.

These demonstrated effects of steroids on transcription provide investigators with the opportunity to study specificity of gene regulation; that is, why do hormones affect only certain cells in the body and then exert their effects on specific genes in those cells? Vertebrates have evolved an elaborate system to transport hormones from the circulation to cell nuclei, where the steroid response is generated. This system also serves to ensure that the hormones will influence only specific organsa necessity, since all cells of the body are exposed to hormones in the circulation. Organs that respond to a particular steroid hormone are called **target organs.** This specificity is due to a difference in the cytoplasms of target and nontarget cells. Target cells are "competent" to respond to steroids because of the presence of specific cytoplasmic receptors that combine with the hormones to form hormone-receptor complexes, which are the functional entities within the target cells that are responsible for the effects of the hormones.

The receptor concept was initially based upon observations that injected radioactive steroids are retained only by hormone target tissues. After injection of radioactive steroid, both cytoplasmic and nuclear binding sites are found in target cells. However, in time, the labeled steroid accumulates in the nucleus. Therefore, steroids reach their site of action by a "two-step" mechanism. *First,* the hormone diffuses into cells from the circulation. In nontarget cells, the hormone is not retained; in target cells, the hormone binds with a cytoplasmic receptor to form a hormonereceptor (H-R) complex, concentrating the hormone in these cells. The cytoplasmic receptors are proteins that have high affinity for specific hormones. A number of receptors have been identified, including those for estrogens, progesterone, androgens, and corticosteroids; the receptors for these hormones are, of course, present only in their respective target tissues. The second *step* is translocation of the H-R complex to the nucleus, where it binds with a nuclear acceptor that has a high affinity for the complex. This results in its accumulation within the nucleus and consequent depletion from the cytoplasm. Nuclear accumulation is also a specific property of target cells, since nontarget nuclei will not acumulate hormone-receptor complexes.

A determination of the specific nuclear site of the steroid acceptors is important in establishing the role of steroids in regulation of cell function. This site has been located by reextraction of radioactive hormone from target cell nuclei. The vast majority of the hormone is discovered bound to chromatin as part of a hormone-receptor complex. These results suggest that the nuclear function of H-R complexes is to influence directly the pattern of transcription on chromatin. Therefore, discovering the identity of the chromatin acceptor and the general mode of interaction of the H-R complexes with chromatin has become a primary objective of research on the steroid response. An important experimental advantage in these investigations is that *isolated chromatin will* bind hormone-receptor complexes. Furthermore, isolated chromatin retains its cell specificity; that is, only target cell chromatin will bind these complexes.

The interaction of H-R complexes with target chromatin allows for the possibility of directly analyzing transcriptional regulation of hormone-responsive genes. Such an analysis for the ovalbumin gene has, been greatly facilitated by the availability of cloned fragments of the gene. A recent investigation with such fragments has demonstrated a 19, base-pair sequence that is present several times in the ovalbumin gene and in its 5' flanking region that binds specifically with the progesteronereceptor complex. Significantly, the sequence is also found upstream from other progesterone-responsive egg white protein genes. Thus, the selectivity of action of the hormone on certain genes within the genome appears to result from an interaction between the'i hormone-receptor complex and sequences flanking those genes that respond. It will be very instructive to learn how this interaction facilitates their transcription.

POSTTRANSCRIPTIONAL CONTROLS

Cellular utilization of structural gene transcripts is not necessarily automatic. The cell is not merely an assembly line that takes any transcript that is synthesized and transports it to the cytoplasm to be used in protein synthesis. This conclusion is obvious when we compare nuclear RNA with messenger RNA. Hybridization analyses have demonstrated that considerably more structural gene transcripts are contained in the nucleus than are associated with polysomes.

The number of different gene transcripts is calculated from the **RNA sequence complexity.** As with DNA complexity RNA complexity is the total length of different sequences, as measured in nucleotides. One method of measuring the complexity of structural gene transcripts is determination of the extent of hybridization of RNA to total single copy DNA, the class of DNA that contains most structural genes. For example, tobacco' leaf cell nuclei have a complexity of 1.19 × 10' nucleotides. These transcripts result from transcription of approximately 19% of the single copy sequences in the genome. However, polysomal RNA (i.e., mRNA) has a complexity of 3.33×10^7 nucleotides, which represents only about 5% of the total single copy sequence complexity of the genome. This fourfold difference between nuclear and polysomal RNA complexity indicates that the majority of nuclear RNA is not represented in polysomal RNA. Similar measurements on sea urchin gastrula RNA have revealed that nuclear RNA has a complexity that exceeds that of polysomal RNA by tenfold. The differences in complexity between nuclear and polysomal RNA suggest that posttranscriptional events may regulate the appearance of some structural gene transcripts on the polysomes.

The necessity of posttranscriptional regulation is also apparent from comparisons of RNA complexity *between* cells of the same organism. These measurements have revealed that the complexity differences between the *nuclear RNAs* of any two cell types are considerably less than the differences between the *polysomal RNAs* of these cells. Additional hybridization studies, using a sea urchin single copy DNA tracer, have revealed that some transcripts that are on polysomes of a given cell type and absent from the mRNA of other cell types are nevertheless present in nuclear RNA of both categories of cells. These results have been interpreted to mean that transcripts may be subject to selective utilization as messengers. This interpretation is supported by recent experiments with specific cloned sequences.

The results of these various hybridization studies indicate that cells must rely very heavily upon posttranscriptional mechanisms to ensure that specific proteins are produced at the right time, in the right cells, and in the correct amounts. The processes involved in utilizing transcripts for the production of functional proteins. These include the series of modifications that RNA must undergo after transcription, the transport of messengers to the cytoplasm, and their utilization by the translational machinery. These processes are subject to regulation that can delay, prevent, or affect the rate of messenger translation. Each of these modes of posttranscriptional regulation is known to be utilized during plant and animal development. Some clearcut examples of posttranscriptional regulation have been elucidated in the literature and will be discussed in this book. Some of these are reviewed on the following pages.

Delayed Translation

Storage of RNA for translation at a later stage of development is one mechanism of posttranscriptional regulation. Very early animal embryos provide an excellent example of delayed mRNA translation. The fertilized egg is involved in very intense protein synthesis. Nuclear function, however, is not necessary for this protein synthesis, since translation continues unabated when the nucleus is removed or transcription is inhibited. These eggs contain the complete protein synthetic machinery in the cytoplasm-including mRNA, tRNA, and ribosomes-for immediate utilization_ following fertilization. The stored messengers are called "masked oogenic messengers"; they are synthesized during oogenesis and stored as mRNP particles until their utilization after the development of the egg begins.

Delayed utilization of transcripts is also typical of organisms that exhibit an interruption in development during early developmental stages. In these organisms, transcripts are stored until development resumes. This mechanism is typical of the early development of seed plants and certain animals. Following plant embryogenesis the seed coat surrounding the embryo hardens, and the embryo dries up, halting development. Typically, the mature seed is shed from the plant, and development does not resume until conditions that are favorable for plant growth and development have been reestablished. After this hiatus, development is reinitiated by **germination.** Protein is synthesized during germination, but synthesis of the RNA templates and the ribosomes for this protein synthesis occurred during embryogenesis, and they were stored in the seed for later utilization. A similar kind of situation is found in pollen. When pollen lands on the style of the flower, a pollen tube may develop. Production of the tube requires protein synthesis, which utilizes stored mRNA and ribosomes.

Dormancy can also occur during postfertilization animal development. For example, the development of the brine shrimp embryo *(Artemia)* can be interrupted at the gastrula stage by rather long dormant periods, during which the embryo becomes dessicated and metabolic functions cease. Dormancy is ended by hydration of these encysted gastrulae, and development resumes, utilizing stored protein synthetic machinery including mRNA.

Differential Longevity of Messengers

In contrast to prokaryotic messengers, which are typically degraded very rapidly after trap scription, eukaryotic messengers may enjoy much longer life spans. This is particularly true for messengers of differentiated cells. Cell differentiation is often accompanied by stabilization in the life span of cell-specific messengers.

Evidence for this differentiation mechanism was first obtained with actinomycin D experiments. Treatment of undifferentiated cells with actinomycin D generally results in rapid inhibition of protein synthesis. However, differentiated cells are much less sensitive to actinomycin D, and protein synthesis in these cells may continue for some time in the absence of RNA synthesis. An additional important observation made during these investigations is that the messengers for cell-specific proteins are particularly stable. When differentiated cells are treated with actinomycin D, the synthesis of cell-specific proteins may continue unabated for some time, whereas the synthesis of other proteins declines rapidly. It appears that one mechanism for narrowing the range of protein synthetic capacities during cell differentiation is differential messenger decay rates. Cell-specific proteins are made on long-lived mRNA templates, and therefore, each template is available for numerous rounds of protein synthesis. However, messengers for nonspecific proteins may be translated only a few times before their decay.

These actinomycin experiments have some major drawbacks, however. Inhibition of RNA synthesis may be incomplete, allowing continued production of mRNA. Furthermore, actinomycin has been found to influence translation as well as transcription.

Messenger stability has been demonstrated by other techniques that do not rely upon inhibitors. One clear-cut example is the globin mRNA of mammalian reticulocytes. The reticulocyte has no nucleus but is active in protein synthesis, utilizing RNA that was synthesized prior to nuclear loss. The pattern of protein synthesis in these enucleated cells might be expected to change as a result of the loss of short-lived messengers during cell maturation. This process would produce an enrichment for longlived messengers, causing protein synthesis to become more homogeneous in time. In fact, the protein synthetic pattern of reticulocytes does exhibit this phenomenon. As the cells mature, the synthesis of nonglobin proteins declines relative to that of globin. The reticulocyte is atypical in that it lacks a nucleus. However, many differentiated cells undergo considerable reduction in RNA synthetic capacity as they mature. Hence, differential messenger decay rates may play a significant role in specialization of protein synthesis in a great variety of cell types. One factor affecting transcript stability might be the 5' cap, since uncapped transcripts may be more susceptible to degradation by certain nuclease enzymes. Thus, regulation of capping might affect protein synthesis indirectly by altering transcript stability.

Regulation of Translation

Utilization of messengers by the translational machinery depends upon successful completion of the initiation step. Regulation of initiation could therefore have a profound effect upon protein synthesis. We shall examine three examples of regulation at this level.

1. The globin messengers of the rabbit reticulocyte provide clear-cut evidence that regulation of initiation can differentially affect the production of specific proteins. Two kinds of globin polypeptide chains are being made in these cells: α and β. The synthesis of these polypeptides is directed by two distinct mRNAs. The rates of initiation of translation of these messengers are different: The aglobin mRNA initiates protein synthesis only 60% as often as β-globin mRNA. Thus, ribosomes are loaded onto the β-globin messenger more rapidly than onto the α-globin messenger. This results in more ribosomes being present per unit length on the β-globin messenger than on the α-globin messenger. Thus, even though both messengers are approximately the same length, the β-globin polysomes average five ribosomes, whereas a-globin is made on polysomes with an average of three ribosomes.

These results clearly demonstrate that initiation can occur at different rates on two kinds of messengers that are simultaneously present in the same cell. Initiation rates could be affected by any mechanism that alters the formation of the initiation complex. For example, messengers might have specialized nucleotide sequences upstream from the AUG codon, which could affect initiation. Another possibility is the existence of messenger-specific initiation factors that could alter the rate of initiation complex formation. There is at present no clear-cut evidence for α- or β-globin mRNA -specific translation factors. Nucleotide sequencing techniques may reveal whether the messenger sequences play a role in initiation.

2. Although messenger-specific translation factors have not been demonstrated for the α-or R-globin messengers, such factors do appear to be involved in modulating the rates of translation in some cells. Evidence for such factors comes from assessing the effects of initiation factors from different cells on cell-free translation of messengers. One example involves experiments utilizing initiation factors of two types of embryonic chick muscle. Both red muscle and white muscle synthesize myosin, but only red muscle synthesizes myoglobin. *In vitro* translation of myosin and myoglobin mRNA reveals differences in translation factors between these two cell types. Myosin mRNA can be translated *in vitro* only when initiation factors from muscle cells are added

to the cell-free translation system. Both red muscle and white muscle have these initiation factors, which is consistent with the ability of both cell types to produce myosin in *vivo*. On the other hand, when myoglobin mRNA is added to a cell-free system, red muscle initiation factors are effective in promoting translation of the messenger, whereas white muscle factors are not. These results suggest that the factor that is necessary for translation of myoglobin mRNA is present only in red muscle. The results do not imply that every messenger has a specific initiation factor that is necessary for its translation. They do suggest, however, that certain differentiated cells have initiation factors that recognize cell-specific messengers and make the utilization of those messengers more efficient in the large-scale production of cell-specific proteins.

3. Mammalian reticulocyte lysates have long been a favored system for in *vitro* studies of translation. Experiments with this system have demonstrated yet another mechanism of regulation of initiation. Protein synthesis in these lysates declines very rapidly unless hemin is added to the incubation mixture. This decline is due to a block in the formation of initiation complexes.

This effect is produced by a translational inhibitor that forms from a latent proinhibitor in the absence of hemin. The presence of hemin greatly retards the conversion of proinhibitor to inhibitor, apparently by interfering with the normal cellular mechanisms that promote this conversion. It remains to be seen whether this mechanism can discriminate for or against specific messengers.

6

The Cellular Basis of Morphogenesis

The cell is the basic unit of morphogenesis. Therefore, an understanding of how cells change shape is important for understanding the processes of morphogenesis. However, morphogenesis is a highly ordered process, and cells change shape in concert with their neighbors, rather than at random. Thus, the consequences of cellular shape changes are highly dependent upon intercellular relationships.

During animal embryogenesis, cells are either tightly bound to their neighbors in a sheet of cells **(epithelial cells)** or exist as individual cells that are free to move about and acquire new locations in the embryo. Virtually all morphogenic processes can be explained by examining the reorganization and rearrangements of epithelia and free cells. The behavior of epithelia depends in turn upon the cells of which they are composed; that is, changes in the form of epithelia occur because of shape changes in their component cells, which cause deformation of the sheet. Conversely, if cells are present as free cells, shape changes are responsible for their individual movement. At the completion of their migratory phase, free cells can form associations with their new neighbors to which they may become tightly bound. Consequently, in addition to changes in shape, the nature of the interaction between cells is an important factor in morphogenesis.

A major distinction between the behaviors of animal and higher plant cells during morphogenesis is the immobility of plant cells due to the presence of cellulose cell walls. The rigidity with which plant cells are maintained in position precludes the dramatic rearrangements of cells that characterize animal development. Hence, the shapes of the bodies of higher plants must be formed by mechanisms that regulate the shapes of immobile cells.

The ultimate regulation of cellular structure and function in both animal and plant development emanates from the expression of the genome, which directs the synthesis of molecules that regulate cellular shape and affect cellular interactions. Contemporary research in morphogenesis is aimed at identifying those molecules and examining their roles in cellular dynamics.

THE CYTOSKELETON

The production of cell shape is mediated by a complex array of fibers in the cytoplasm that is variously referred to as the **cytoskeleton** or the **cytoplasmic fiber system.** The term cytoskeleton defines the role of the fibers in maintaining cell shape but it has the unfortunate connotation of rigidity and permanence. On the contrary, the cytoskeleton constitutes an exceedingly dynamic system that is capable of considerable change and provides the means whereby the cells of the embryo can change their shapes and thereby produce the form of the adult organism.

In a recent review of the cytoskeleton, Cohen (1989) quoted from a 1733 essay by Alexander Pope to emphasize that this complex system of fibers has an underlying structural basis that functions in cellular morphogenesis. Pope wrote: "A mightly maze! but not without a plan." In this section we shall attempt to untangle the maze and demonstrate how it functions in cellular morphogenesis. The cytoskeleton is composed of **microtubules, microfilaments,** and **intermediate filaments**. Cellular shape change is brought about primarily by the microtubules and microfilaments.

Microtubules

Microtubules are present in all cells at some point in their life cycles. They are important functional elements in the mitotic spindle, are involved in translocation of cellular organelles from one location to another in 'the cytoplasm, and are responsible for the movement of cilia and flagella. But they play a very specific role in generation and maintenance of cell asymmetry. In asymmetrical cells they are often aligned in parallel bundles in the cellular long axis. In later chapters we shall discuss several examples of cellular asymmetry that is produced by microtubules, including amphibian neural ectoderm cells and the filopodia of sea urchin primary mesenchyme cells.

Microtubules are hollow cylindrical rods, 25 nm in diameter, formed of 13 rows of **protofilaments** that run parallel to the microtubule long axis. The protofilaments are composed of two similar kinds of protein subunits, called α- and β-tubulins. *In vitro* studies of microtubules have been very instructive in revealing how the tubulins are assembled into microtubules. One molecule of u.-tubulin combines with one molecule of β-tubulin to form a **tubulin dimer.** The dimers, in turn, polymerize end to end to form the protofilaments. Concomitant side-by-side assembly of the 13 protofilaments produces a microtubule. Each microtubule is polarized such that all the dimers are oriented in the same direction with respect to their α- and R-tubulin subunits. This mode of assembly is called head-to-tail assembly.

In vitro microtubule assembly will occur rapidly at 37°C if sufficient subunits are present, but depolarization occurs at low temperatures. Other factors involved in microtubule assembly are the nucleotide GTP and the divalent cation Mg^{++}. Addition and loss of dimers can only occur at the ends of microtubules. Margolis and Wilson (1998) demonstrated that assembly and disassembly occur at both ends of microtubules under equilibrium conditions *in vitro*. However, these opposing events occur at different rates at the two ends; that is, one end favors subunit addition (the assembly end), and the other end favors subunit loss (the disassembly end). A consequence of this mode of polymerization is tubulin flux; that is, subunits within the microtubule are continually displaced toward the disassembly end where they are eventually lost. This phenomenon is called treadmilling.

The results of *in vitro* studies have important implications for in *vivo* behavior of microtubules and their regulation. Clearly, elongation of microtubules can occur only by addition of dimers to the ends of microtubules and only if the equilibrium favors assembly over disassembly. Conversely, microtubules will shorten if

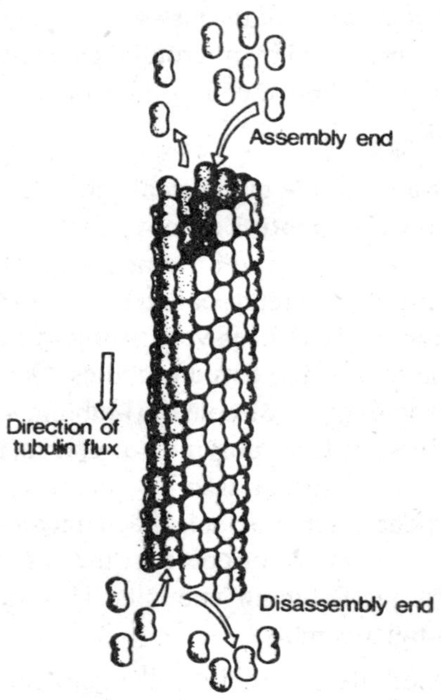

Figure 6.1 : Microtubule assembly and disassembly. Tubulin dimers are added and lost at both ends of the microtubule under equilibrium conditions in vitro. At the assembly end, the rate of subunit addition exceeds the rate of subunit loss. Subunits added at this end treadmill through the microtubule and are lost at the opposite end. Note that a microtubule is composed of 13 rows of protofilaments organized into a hollow cylinder.

disassembly is favored. Although polymerization of dimers will occur spontaneously in *vitro* by addition of the correct components at a favorable temperature, microtubule assembly *in vivo* is a highly regulated process, with both the length and orientation of microtubules under close control. A major difference between *in vivo* and in *vitro* microtubule assembly is that there are no free microtubules in cells. All cellular microtubules have one end embedded in a structure called the **microtubule organizing center (MTOC),** which corresponds to the centrosome and contains centrioles and surrounding pericentriolar material. The MTOC serves as the point of initiation of growth for microtubules and may specify the number, distribution, and length of microtubules that emerge. Once microtubules are initiated, their continued growth occurs by tubulin dimer addition to their distal tips. A critical level of available tubulin subunits is a prerequisite for growth. Other factors that may be involved in control of microtubule growth are uncertain. One possibility is that **microtubule-associated proteins (MAPs),** which have been shown to promote microtubule assembly in *vitro,* are involved in growth control. These proteins associate with the surface of the microtubule after dimers are incorporated into the growing polymer. The MAPS stabilize the polymer and thus shift the equilibrium of the polymerization reaction toward assembly and away from disassembly. However, it remains to be seen whether MAPs play this same role or, indeed, any functional role in regulating microtubule growth in the intact cell.

The occurrence of treadmilling *by in vivo* microtubules is as yet uncertain. One reason for the uncertainty is that it has yet to be deter mined whether cellular microtubules are free to exchange subunits at both ends as occurs *in vitro;* the anchoring of microtubules at one end: in the MTOC may preclude this. If treadmilling does occur, it is easy ti, visualize how it could be used by cells to transport materials from place to place in the cell. Cellular components that are attached to treadmilling subunits would be displaced toward the disassembly end of a microtubule. Such a mechanism might be involved, for example, in chromosome movement on the spindle during mitosis.

The functional properties of microtubules in morphogenesis are due primarily to the cell's ability to elongate the microtubules rapidly, to maintain them once they are formed, and to shorten them when necessary. Drugs that affect microtubules are important tools for determine whether microtubules are involved in a particular morphogenic process. Colchicine is the most commonly used drug. It has been used for a number of years as a reversible inhibitor of mitosis. We now know that its effect on mitosis is due to its binding to tubulin subunits, which upon incorporation into a microtubule prevent additional polymerization. A relatively new synthetic antimicrotubule drug called **nocodazole** has recently been added to the developmental biologist's arsenal. This drug also binds to tubulin subunits and inhibits polymerization of microtubules. When either col: chicine or nocodazole is applied to developing systems, it serves as an excellent probe for determining whether microtubules are involved in a morphogenic process. An alternative to the use of drugs is low temper ature, which causes reversible depolymerization of microtubules.

Because microtubules play a key role in shaping cells, their orien· tation in the cytoplasm must be precisely regulated. The microtubule organizing centers may affect the orientation of microtubules; if so, it still would be necessary to account for the orientation of the centers themselves in explaining how orientation is regulated. Furthermore, the formation of organized structures during development requires that the shapes of individual cells in a multicellular array be coordinated with, one another. Hence, it would also be necessary to account for the co-ordinated orientation of organizing centers of contiguous cells. Clearly, we are only just beginning to understand the regulation of microtubules,

Microfilaments

Microfilaments are also universal animal cell constituents, as shown by their roles in the separation of daughter cells at cytokinesis and in cytoplasmic streaming in plants. In morphogenesis they are, utilized in cells to cause localized contractions, such as the shortening of pseudopodia or the narrowing of cell diameters. Microfilaments can be resolved with the electron microscope as 6-nm diameter threads. Contractile microfilaments are usually seen just below the cell membrane. They make functional contact with the membrane, which would account for their ability to cause cellular deformations when they contract. Microfilaments are present as close parallel arrays. These arrays can assume a number of configurations: bundles, sheaths, or meshes, networks, or lattices. None of these configurations are considered permanent;

the microfilament system is capable of rapid rearrangement as well as local assembly and disassembly.

The configuration of microfilaments has been correlated with cellular activity of tissue culture cells. Cells that are nonmotile and stretched on the substrate have microfilaments that are organized into stress fibers. The microfilaments in this configuration may be primarily structural, assisting in maintaining cell shape. In motile cells, a few microfilament bundles are seen. However, most microfilaments are in a diffuse network. These microfilaments are considered to be primarily contractile. Two lines of evidence indicate that microfilaments are composed of the contractile protein actin, which is also one of the contractile proteins in muscle: (1) Microfilaments interact with heavy meromyosin (HMM) to form complexes that are visible with the electron microscope. HMM is a fragment of myosin that is produced by proteolytic cleavage of the larger molecule. (2) Antibodies to actin will bind to microfilaments.

Like microtubules, microfilaments are polymers, in this case formed from actin subunits. The polymers are frequently called F-actin (for filamentous actin), whereas the unpolymerized subunits are G-actin (for globular actin). This sort of organization imparts dynamic properties to the actin network, which allow it to adapt rapidly to the changing requirements of cells during morphogenesis. As with tubulin polymer ization, *in vitro* polymerization of actin subunits occurs in a head-to-tail; fashion onto both ends of microfilaments. Polymerization is favored by the presence of Mg^{++}. The G-actin molecules add more readily to one end (the assembly end) of the microfilament than to the other. At equilibrium, subunit loss at the disassembly end is balanced by subunit addition at the other end, and filament length remains constant. Actin subunits would, therefore, treadmill through the filaments. Hydrolysis of ATP is necessary to maintain actin monomer treadmilling. Growth of microfilaments occurs when the rate of subunit addition exceeds the rate of subunit loss.

Microfilaments *in vivo* emerge from sites on membranes. Elongation apparently occurs by addition of subunits at these membrane-associated nucleating sites and not at the free, cytoplasmic tips. Thus, the direction of microfilament growth is opposite that of microtubules, in which subunits are added at the free distal tips rather than at the ends that are embedded in the MTOCs. As previously discussed, the association of microfilaments with the cell membrane accounts for their ability to deform the membrane when they contract. The membrane association is thought to be mediated by a protein called **vineuiin,** which has been localized to sites of microfilament-membrane association.

The generation of contractile forces by microfilaments is based upon an actin-myosin sliding-filament mechanism that is similar to that in smooth muscle. Low levels of myosin are found to be associated with microfilaments in nonmuscle cells. Contraction is regulated by Ca^{++} interaction with the calcium-binding protein, calmodulin. Calcium causes conformational changes in calmodulin, which can alter its biological function. A rise in the concentration of free Ca^{++} results in formation of the Ca^{++}-calmodulin complex. Calcium-bound calmodulin activates a cyclic AMP-dependent kinase enzyme, which phosphorylates the myosin light chain. This, in turn, leads to activation of myosin ATPase by actin, and contractility ensues.

Two other proteins that are components of the contractile mechanism of muscle may also be found associated with microfilament bundles. These are **tropomyosin** and α-actinin. They may provide interfilament crossbridges within the bundles. Alpha-actinin is particularly intriguing because its ability to cross-link filaments is regulated by calcium: High Ca^{++} levels cause the cross-links to break, whereas low Ca^{++} levels allow them toreform rapidly. The calcium sensitivity of α-actinin may help facilitate rapid change in the cellular microfilament network: Cross-linking of microfilaments would occur under low calcium conditions, and a calcium flux would release the filaments to allow their rapid reorganization. Tropomyosin is found primarily in association with microfilaments in bundles of immotile cells but not in association with microfilaments of highly motile cells. This correlation suggests that tropomyosin has a stabilizing influence on structural microfilaments, rather than an involvement in contraction.

Numerous other proteins have also been reported to associate with actin microfilaments in various cell types. These actin-binding proteins fall into three functional categories: (1) proteins that cross-link microfilaments to form bundles; (2) proteins that modulate actin polymerization; and (3) proteins that affect microfilament contractility.

The microfilament network can be altered experimentally by exposing cells to **cytochalasin B,** a drug that blocks actin polymerization. Thus, it can be used to determine whether microfilaments are involved in producing particular shape changes. Unfortunately, the effects of cytochalasin on the cell are not so specific as those of colchicine; it causes side effects, including alterations of membrane transport functions. Consequently, when evaluating the effects of cytochalasin, the investigator must consider the possibility of nonspecific side effects.

Intermediate Filaments

The third category of cytoskeletal components is the intermediate filament. Intermediate filaments have a mean diameter of 10 nm. Unlike microtubules and microfilaments, intermediate filaments are not composed of the same molecular component in all cell types. Instead, the intermediate filament composition varies according to cell type and is dependent upon cell differentiation. The subunit structure varies among the five major classes of intermediate filaments. The five classes and the cell types in which they are found are: (1) keratin filaments (epithelial cells); (2) neurofilaments (neurons); (3) glial filaments (glial cells); (4) vimentin filaments (most differentiating cells, tissue culture cells, and in certain fully differentiated cells); and (5) desmin filaments (muscle and certain nonmuscle cells). Filaments of different types can co-exist in the same cell. For example, both keratin and vimentin filaments have been found in cultured epithelial cells.

It has been proposed recently that intermediate filaments interact with cellular organelles so as to help organize and maintain their threedimensional arrangement in the cytoplasm. Vimentin filaments are closely associated with the nucleus and during mitosis form a cage that surrounds the mitotic spindle. During mitosis the nuclear envelope, cytoplasmic microtubules, and microfilament bundles break down. Perhaps the vimentin filaments, which are the only cytoskeletal elements retained throughout mitosis, provide the framework that helps to reconstruct cytoplasmic geometry after division.

THE CELL SURFACE

The cell surface is the region of a cell that interfaces with other cells, Consequently, the surface must mediate and integrate cellular activities that involve cell-cell associations. Since cellular interactions are dynamic, it follows that cell surface properties are also dynamic. The surface also must be able to accommodate changes in its configuration under various circumstances. For example, it must be able to expand to allow cell growth. It also must be capable of fusion with other membranes (e.g., fertilization or fusion of the plasmalemma with intracellular vestcles) and can accommodate removal of a portion of itself (e.g., formation of vesicles during endocytosis). These dynamic properties of the cell surface are due to its molecular composition and organization.

The current working model for the cell surface is based upon the **fluid mosaic membrane model** of Singer and Nicolson. According to this model, the cell membrane is composed of a bilayer of phospholipids into which numerous proteins are inserted. The lipids are arranged as a double row of molecules whose hydrophilic heads are o the bilayer's outer face and whose hydrophobic tails are pointed toward the center of the bilayer. The membrane proteins are of two types, based upon the extent of their interaction with the membrane lipids. The **integral** (intrinsic) **membrane proteins** penetrate both the hydrophilic and hydrophobic zones of the membrane and are tightly bound to the lipid phase, located primarily in the hydrophilic zones. Some of these proteins protrude from both the inner and outer faces of the membrane, whereas others are intercalated into either the inner or outer face. The proteins that protrude from the outer face are usually covalently bound to car' bohydrates and are, therefore, **glycoproteins.** As we shall see, many of the morphogenic properties of the cell surface involve these carbohydrates which mediate interactions with other cells, such as cell-cell adhesion The proteins that protrude from the membrane into the cytoplasm c interact with cytoplasmic constituents, such as elements of the cytoskeleton.

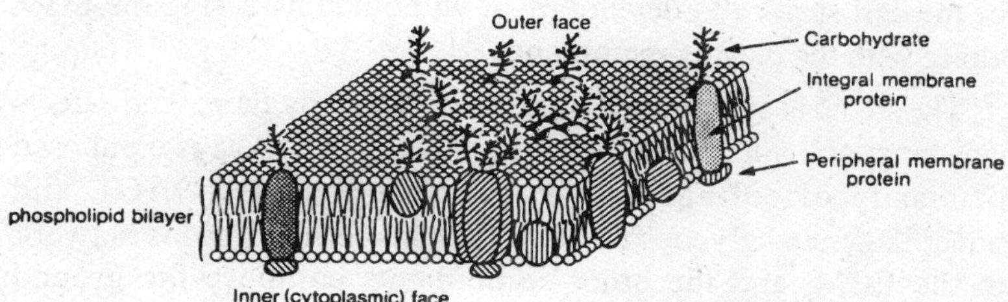

Figure 6.2 : Fluid mosaic membrane model of Singer and Nicolson.

The other category of membrane proteins (most of which are also glycoproteins) constitutes the **peripheral** (extrinsic) **membrane proteins.** These proteins are bound to either the inner or outer face of the membrane via weak bonds with either the lipids or the integral membrane proteins. On the outer surface they could mediate cellular interactions. On the inner surface some of them help to mediate the attachment of the cytoskeleton to the integral membrane proteins, whereas others serve to stabilize membrane structure by restricting mobility of the lipids and intrinsic membrane proteins.

The spatial arrangement of membrane components is dynamic; that is, they are laterally mobile in the membrane. Rearrangements of membrane components will allow for spatial alterations in membrane properties and can indirectly affect cytoplasmic structure via the linkages between membrane proteins and the cytoskeleton. Since the lateral mobility of membrane components can have dramatic effects on membrane properties, order and stability must be maintained in the membrane except when change is required. Mobility is restricted by peripheral membrane proteins or by binding of membrane proteins to the cytoskeleton or to components of the extracellular matrix. Likewise, rearrangements of membrane components may be caused by a change in their interaction with these intracellular and extracellular elements.

The extracellular matrix (ECM) is the material that surrounds connective tissue (i.e., free) cells and underlies epithelia. Molecules of ECM interact with the cell membrane in such an intimate way that the membrane and the molecules form a functional and structural continuum. Hence, it is impossible to consider the role of the cell surface in development without considering the ECM along with the plasma membrane.

The ECM is synthesized by cells, secreted by them, and affects cell function by its interactions with them. The ECM is composed primarily of **collagens, glycosaminoglycans** (GAG), and noncollagenous glycoproteins. Collagens form a fibrous network in the ECM, and the other components compose the ground substance that is intercalated among the collagen fibrils. GAG consist of long, unbranching chains of repeating disaccharides, one of which is a hexosamine, which may be sulfated, and the second is hexuronate or D-galactose. The glycosidic bond between these units may be in the β- or α-configuration. With the exception of hyaluronic acid, GAG (e.g., chondroitin, chondroitin sulfate, dermatan sulfate, keratan sulfate, heparan sulfate, and heparin) are linked to proteins to form **proteoglycans.** A proteoglycan molecule contains a long core protein from which numerous GAG chains extend. Variations in the number, type, and distribution of GAG on the core protein as well as variations in the core protein itself can generate a great many different proteoglycans.

Hyaluronic acid serves to link together individual proteoglycans to form an extensive network that associates with collagen in the ECM, The physical properties of the collagen in the matrix can be modified by certain proteoglycans. Hence, the structure of the matrix may exhibit' local specialization and undergo changes during development that are controlled by the proteoglycan composition. These variations in the matrix may affect such cellular functions as their migratory properties, since cells migrate through the ECM, which they use as a scaffold. GAG also have been shown to interact directly with ce surface molecules. Such interactions may influence cell differentiation by altering various parameters of cell function.

A variety of glycoproteins are constituents of the ECM. One

of the best studied of these is fibronectin, which is a high molecular weigh molecule that is widely distributed in connective tissue, at cell surfaces, and in **basal laminae** (the collagenous zones that underly epithelia and surround muscle cells). Fibronectin has binding sites for cell surface receptors as well as collagen and GAG and may be involved, therefore, in mediating cellular interactions with these components of the extracellular matrix. Another glycoprotein, laminin, which is more limited in distribution, is found primarily in basal laminae an is involved in attachment of epithelia to collagen.

THE CELL SURFACE AND CYTOSKELETON IN MORPHOGENESIS OF ANIMAL EMBRYOS

Morphogenesis involves a number of distinct cellular functions; the embryo forms as a consequence of the integration and control of these processes in the separate cells of which it is composed. Research in morphogenesis is aimed at identifying these cellular functions, understanding their molecular basis, and clarifying the mechanisms that control and integrate cellular morphogenic functions to produce the organism. The formation of a distinct, recognizable organ during development depends upon the mobilization of the cells that will form it. In most cases those› cells arise at some distance from where the organ will be formed. Theymust dissociate from their original position, migrate through the embryonic connective tissue, recognize their correct new position, and become fixed in these new positions. After fixation, the individual cells of aggregate change shapes and undergo differentiation to become fun tional elements of the forming organ. In this section we shall disc the roles of the cell surface and the cytoskeleton in the changing relationships among cells during development. The role of the cytoskelet, *in* shaping *immotile cells* in the formation of tissues and organs.

Cell Motility

Cell motility plays a key role in cellular reorganizations that are necessary in morphogenesis of the animal embryo. We shall discuss several examples of migratory cells in later chapters. These will include: (1) sea urchin mesenchyme; (2) inward moving cells of vertebrate gastrulae; (3) vertebrate neural crest cells; (4) primordial germ cells of vertebrate embryos, which originate in endoderm and migrate to presumptive gonads; and (5) vertebrate mesenchyme cells, which migrate within the organizing embryo and condense around epithelia. In most of these cases, analysis of the mechanisms of migration is difficult because the cells are inaccessible to observation. For this reason, much of our knowledge of cell motility is derived from studies of vertebrate fibroblast cells in tissue culture. Although movement of fibroblasts in *vitro* is somewhat artificial, it provides a useful model for analyzing *in vivo* cell movement.

Fibroblasts move by spreading along a substrate, with which they form temporary adhesions at several points. The forward movement of the cell occurs by the protrusion of the leading edge of the cell to form a thin, fanlike structure called the **leading lamella.** After protruding forward, the leading lamella attaches to the substrate near its leading edge at points of adhesion, called **focal contacts.** As the leading edge thrusts forward, the cell elongates in that direction because the trailing edge retains its attachment for a time. The breaking of the adhesion at the trailing edge occurs as the cell body is retracted up to the point of the next adhesion, and the cell is displaced forward. Thus, continuation of movement depends upon a repeated process of forward thrust of the leading edge, adhesion of the leading edge to the substrate, contraction of the cell, and loss of adhesion at the trailing edge. The adhesion points themselves do not move; as old adhesions fade, they are replaced by new focal contacts that are formed near the leading edge of the lamella in front of the existing ones.

The motile force for migrating fibroblasts is provided by microfilaments. Microfilaments have two distinct kinds of distributions in these cells. The leading lamellae are flattened projections of the cell cortex containing a lattice-work of microfilaments between the upper and lower cell membranes. In time-lapse cinemicrography the lamellae are seen to be highly active structures, going through cycles of random protrusion and withdrawal. Forward movement occurs because protrusion is more frequent and lasts longer than the withdrawal phases. The anterior margin of the lamella also undulates by an upfolding that lifts a portion of the margin above the substrate. This undulating activity is called "ruffling". Since the activities of the leading edge are frozen by cytochalasin B, it is likely that the activities of the lamella are produced by the microfilament network that is located there.

A quite different organization of microfilaments is found in the bundles, or cables, of microfilaments that are anchored on the cytoplasmic faces of the focal contacts, presumably by vinculin. These bundles are oriented parallel to the direction of movement and project backward to merge into fibrillar material that surrounds the nucleus. They are thought to be responsible for drawing the posterior end of the fibroblast forward by their contraction. As the microfilament bundles contract, the front edge of the cell extends forward and lays down new focal contacts, which acquire new microfilament bundles. Thus, with the contractile forces of a succession of microfilament bundles from progressively more anterior positions, the cell is pulled forward. The contractile function of the microfilament bundles is supported by observations that cytochalasin B inhibits fibroblast locomotion comotion.

Microtubules are thought to play a structural role in migrating cells, Microtubules are oriented parallel to the long axis of migrating fibroblasts. They are considered to be responsible for the elongation of the cells, an interpretation that is supported by the rounding up of fibroblasts in the presence of colchicine.

The protrusive activity of the leading edge of a migrating fibroblast is inhibited by adhesion to another fibroblast. This phenomenon is called **contact inhibition of migration**. Once a leading lamella is suppressed by making contact with another cell, protrusive activity is initiated elsewhere, and a new front end is formed as the cell changes its direction of migration. If cell density is sufficiently high, cells contact one another on all sides and are unable to migrate further. The result is the formation of a confluent monolayer of cells that are inhibited from movement by their mutual contacts. Cells in a confluent monolayer have a reduced rate of cell division. This is called contact inhibition of mitosis.

The movement of tissue culture cells is not necessarily the same as that of embryonic cells migrating within the embryo. Individually migrating cells *in vivo* are not flattened with thin leading lamellae. Instead, they appear to move by means of narrow pseudopodia and blunt protrusions. These differences probably arise because cells *in vitro* are migrating along a flat inanimate substrate, whereas cells *in vivo* migrate through the fibrous extracellular matrix. It was long ago pointed out by Harrison (1984) that cells assume different shapes *in vitro* according to the mechanical conditions to which they are exposed. Weiss (1999) stated that the shape of a migrating cell is "the result of a distinctive behavioral reaction of a living cell to its environment." Hence, the dif. ferent shapes of cells migrating *in vivo* and *in vitro* may be of secondary importance.

Cell migration in embryogenesis is highly precise. Groups of pre cursor cells migrate in streams from their origin to their target sites. For example, the primordial germ cells migrate from the endoderm to the genital ridges and populate them to become gametes. Also, neural crest cells migrate from their middorsal position to form visceral cartilage, cartilage of the embryonic skull, adrenal medulla, spinal ganglia, and other structures. The mechanisms that direct migratory cell populations toward their target sites are very important for the proper spatial organization of the embryo.

Weiss (1999) demonstrated that cells tend to follow fibrous pathways during their migration. This phenomenon is called contact guidance. Dunn and Ebendal (1998) observed that a three-dimensional mesh. work of fibers is more effective in guiding cells than is a flattened mat of fibers. Since embryonic cells migrate through a three-dimensional meshwork of fibers in the extracellular matrix, it has been proposed that this meshwork guides the cells toward their intended sites. Perhaps local variations in the matrix, such as glycosaminoglycan heterogeneity can affect the shape of the matrix, and hence, the path of migration. Othe factors that could affect the path of migration are: (1) the nature of th adhesions that the cell forms with the substrate; (2) physical barriers that block the migratory path; and (3) molecular substances that attract cells (chemotaxis). The latter mechanism is utilized to orient migrating slime mold cells.

Since cellular interactions with the substrate are so important in cell motility, we shall now examine that relationship in more detail. Cell membrane glycoproteins function as receptors that bind to molecules of the extracellular matrix. One of the most important components of the matrix is fibronectin, which has been demonstrated to promote cell motility *in vitro*, possibly because fibronectin is involved in formation of the focal contacts with the substrate and intracellular organization of the bundles of contractile elements that emanate from these sites. Its role in cell adhesion may be explained by the fact that it can bind both to cell membrane receptors and to components of the extracellular matrix such as collagen and glycosaminoglycans. Thus, it could serve to link a cell to the extracellular matrix. Fibronectin forms extracellular fibrils that are organized parallel to the long axes of leading lamellae. Therefore, it is possible that fibronectin fibrils influence the direction of migration. This might occur by affecting the organization of the microfilament bundles originating at the focal contacts. The microfilament bundles may give directionality to cell movement by serving as guides along which the nucleus and the bulk of the cytoplasm move during forward displacement.

Although fibronectin promotes motility of fibroblasts *in vitro,* it is uncertain whether it performs this function in the embryo. Fibronectin has been detected in the extracellular matrix along pathways of cell migration in various embryos. Thus, it is present at the right time and place to promote *in vivo* cell migration. Further investigation is necessary to establish a functional role for fibronectin in this process.

Intercellular Adhesion

One of the most important factors governing the behavior of an individual cell of the embryo and the role of that cell in morphogenesis is the nature of its interactions with its neighboring cells. The animal embryo takes shape as its cells sort out and associate into specific multicellular assemblages that are the primordia for the organs and tissues of the adult organism. During this period, cells change their relative positions as a result of modifications in the adhesive interactions between cells. For example, adhesion between cells maintains the integrity of epithelial layers, and conversely, reduced adhesion allows cells to dissociate from an epithelium and form free cells. The latter may migrate from their site of origin to a distant location where they reassociate to form once again a cohesive unit. Condensation of free cells frequently occurs around an epithelial vesicle, bringing cells of diverse origin together into organ primordia. The assembly of such organ primordia involves cell recognition and cell adhesion. The dynamics of intercellular contacts are, therefore, highly significant for organizing cells into a definitive embryo.

The specialized adhesive properties of cells are demonstrable by experiments in which tissues are dissociated and allowed to reassemble into aggregates. H.V. Wilson (1987) was the first to perform this type of experiment. Wilson dissociated adult sponges into a suspension of cell fragments, individual cells, and small clusters. The cells sink to the bottom of the culture dish, where they actively migrate until they adhere to one another to form clusters that increase in size by addition of individual cells and by fusion with adjacent clusters. The aggregates even.. tually produce small, individual sponges. Specificity of sponge cell ag gregation was demonstrated by mixing cells obtained from dissociate sponges of different species that have cells that are distinguishable by their pigmentation. These mixed cell suspensions at first associate a random, but the cells later sort out as separate aggregates, each contain ing cells of only one species. The cells sort out into species-specific aggregates because cells form more stable contacts with cells of the same species than they do with cells of another species. Thus, sponge cells have a selective affinity for cells of their own species.

Similar experiments have demonstrated that cells of developing vertebrate embryos will segregate in a mixed aggregate according to typ and reestablish their former associations. These studies were initiate by Johannes Holtfreter and his associates with amphibian embryos. Townes and Holtfreter presented the first critical evidence tha embryonic cells with distinct developmental fates have different adhe sive properties. They took advantage of the size and pigmentation differences of ectoderm, mesoderm, and endoderm cells to follow the cells behaviour in aggregates. The cells reorganize within aggregates to assume positions that correspond to the normal germ layer arrangement in the intact embryo. Thus, in a mixed aggregate of mesoderm and presumptive epidermis the mesoderm sinks to the inside to form mesenchyme whereas the epidermal cells accumulate at the periphery and spread over the surface to form an epithelium. When cells of all three germ layers are combined, the endodermal cells form a compact ball whereas cells of presumptive epidermis form a surface epithelium, an mesoderm produces an intermediate mass of **mesenchyme,** which is term applied to free embryonic cells and usually refers to cells of mesodermal origin. Even cells of the same germ layer with different developmental fates will segregate from one another. Thus, presumptive neural ectoderm and epidermis in a mixed aggregate will sort out, with the epidermis forming a surface epithelium and the neural ectoderm sinking to the interior to form structures resembling brain vesisles.

The cell-sorting experiments demonstrate nicely that selective cell adhesion is an important force in embryonic organization. Two important aspects of cellular segregation are evident from these experiment. One is the preferential association of like cells. This undoubtedly plays a role in maintaining the integrity of embryonic cell layers. Likewise, changes in adhesiveness will allow cells from a single layer to form separate cell groupings. The other aspect is the spatial arrangement of cell groups with respect to one another. Thus, during gastrulation, endoderm sinks to the inside of the embryo, while ectoderm spreads over the periphery.

The relative positioning of tissues within an aggregate has also been studied in cell-sorting experiments by using cells from the organ rudiments of chick embryos, which have been dissociated by trypsinization. Dissociated cells of different origins combine to form a common aggregate and then sort out according to tissue type, usually with one tissue surrounding the other. The advantage of using organ primordia is that the tissues formed after cell-sorting will differentiate to reveal their developmental fate. According to Steinberg, the relative positions of cells in a mixed aggregate result from random motility of the cells and quantitative differences in the adhesiveness between them. Thus, cells with the stronger mutual attraction will aggregate in the center, whereas cells with weaker attraction will remain at the surface This is the **differential cellular adhesiveness** hypothesis.

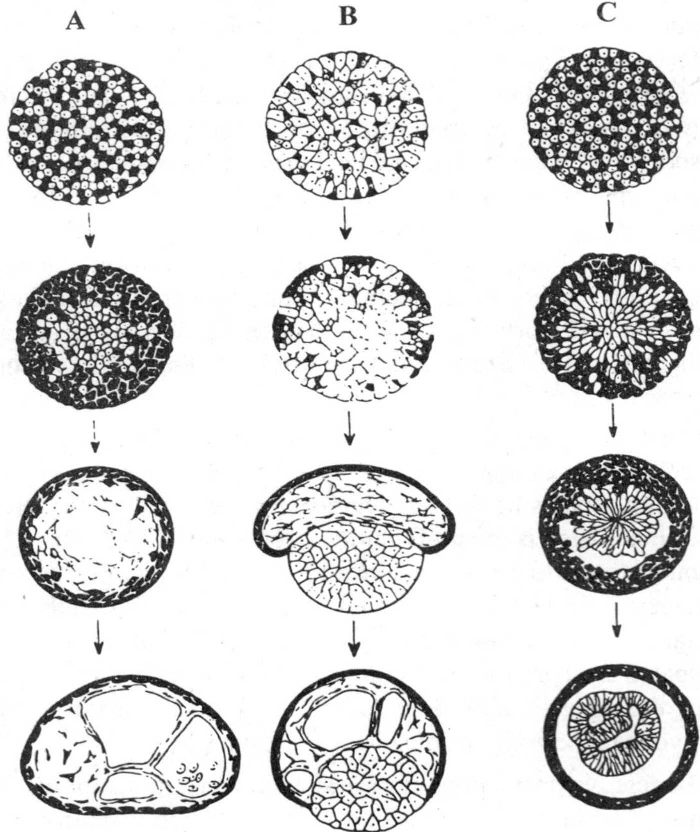

Figure 6.3 : Rearrangement of disaggregated and reaggregated embryonic cells. A, Combined epidermal and mesodermal cells. B, Combined epidermal, mesodermal, and endodermal cells. C, Combined epidermal and neural plate cells.

Consider the results of the following tissue combinations. In a mixture of limb-bud cartilage and heart, the heart cells surround the cartilage. When heart and liver cells are combined, the liver cells surround the heart cells. Finally, in a mixture of liver cells and cartilage cells the liver surrounds the cartilage. Steinberg has proposed that the relationship between liver and cartilage is predictable from the previous two combinations. Heart is less adhesive than cartilage, and liver is less adhesive than heart; therefore, liver should envelop cartilage. Thus, a hierarchy can be arranged whereby those at the top of the scale will segregate internal to those immediately below, whereas those at the bottom of the scale segregate external to all others.

Differential adhesiveness provides a probable explanation for thev ordering of the germ layers during amphibian gastrulation. It However, Steinberg (1990) has emphasized that differential adhesiveness does not completely explain cellular arrangement during the organization of cells into tissues and organs and its maintenance in the adult. Furthermore, the hypothesis does not deal with the chemical or physical mechanisms of cell-cell adhesion. It merely explains the consequences of differences in the strengths of adhesion between various kinds of cells.

Moscona (1980) has proposed that specific cellular **adhesiveness** is an important factor that determines cellular associations. According to hypothesis, cell adhesion is a property of specific cell surface macromolecules that allow cells to recognize like cells and bond with them. Thus, cells belonging to different tissues will segregate from one another in a mixed aggregate because they recognize and preferentially associate with their own kind. This mechanism could also function to cause cells of different origins to come together to form organ primordia. Moscona has proposed that cell-surface constituents function as **cell-cell recognition sites** and **cell-cell** ligands that bridge between specific cells, and thus hold them together. Specificity of cell recognition would be determined by the molecular characteristics of the receptors and ligands and also by their topographical arrangements on the cell surface.

Cell Surface Molecules in Adhesion

The proposal that there are specific adhesion factors on cell surfaces has prompted extensive studies aimed at discovering whether such factors regulating cellular associations can be demonstrated. Cell surface aggregation factors were first detected by analyzing the supernatant fluid in which cell dissociation has occurred. The rationale for these experiments was the assumption that dissociation releases the cells ligands into the medium. Consequently, under conditions that are favorable for reassociation, the addition of the supernatant medium would cause cells to aggregate.

Sponge cell aggregation was examined in just this way. Dissociation of sponge cells occurs in the absence of the divalent cations Ca^{++} and Mg^{++}. Thus, sponges placed in Ca^{++} and Mg^{++} -free seawater will dissociate. The presence of an aggregation factor in the dissociation medium is demonstrated by adding the medium back to dissociated cells, along with the divalent cations. The factor also accounts for the species-specificity of reaggregation, since it will cause adhesion only of cells of the same species. An additional point about these experiments should be made. That is, the cells are maintained at low temperatures at all times to inhibit metabolic activity. If the temperature is raised, the cells will aggregate without added supernatant. This implies that at the higher temperature the cells are synthesizing a new aggregation factor to replace that lost during dissociation.

Aggregation in sponges is a two-step process. First, the extracellular aggregation factor binds to specific cell surface receptors. Second, the aggregation factors bound to adjacent cells interact with one another to link the cells together. The second step requires calcium ions, which explains why aggregation can be controlled by the amount of Ca^{++}. The properties of the aggregation factor are those of an intercellular ligand.

Similar experiments have been conducted to examine adhesion between vertebrate embryonic cells. Chick neural retina cells can then be dissociated with trypsin; these cells can then maintained in culture, where they will reaggregate to reconstruct retinal tissue. When mixed with other chick embryo cells, the cells sort out to form tissue-specific aggregates. As with sponge cell aggregation, chick retina cells will not reaggregate when they are maintained at low temperatures. Inhibitors of macromolecular synthesis will also prevent reaggregation. Thus, metabolic activity is required, presumably to resynthesize the ligand.

When retina cells are maintained in culture as a monolayer, the ligand produced by the cells apparently accumulates in the culture medium, since the medium can be used to promote retina-specific cell aggregation. This retina-specific factor has been purified and characterized as a glycoprotein. It is called retina cognin to indicate its presumed role in recognition and association of embryonic retina cells. Retina cognin has been isolated directly from the membranes of embryonic neural retinal tissue, indicating that it is indeed component of the cell surface. Another important property of the retina specific factor is that it is produced only by retina cells from pre-13-day embryos. This suggests that it is developmentally regulated and produced by retina cells when it is needed for a specific developmental even. Antiserum to retina cognin has also bee used as a probe to detect cognin and to investigate its functional role. The surfaces of chick embryo retina cells treated with rabbit anticognin followed by fluorescent goat antirabbit serum, for discussion of this technique) fluorescent in ultraviolet light. Thus, the anticognin recognizes the cognin on the cell surfaces. Further more, treatment of retina cells with anticognin abolishes the ability, the cells to aggregate. Clearly, this surface-associated molecule is involved in retina cell aggregation.

Chick neural retina cell adhesion has been the subject of numerous investigations. These studies have revealed a number of cell surface molecules involved in adhesion of these cells. One example is **cell adhesion molecule** (CAM), which has been studied intensively with anti-CAM antibodies. Another series of investigations have revealed the existence of two retina cell adhesion mechanisms that can coexist in the same cell. One of these mechanisms requires ionic calcium for adhesion, whereas the other is calcium independent. Studies of the possible involvement of cognin, CAM, and other adhesion molecules in these two mechanisms should help to clarify the total number of adhesion mechanisms for neural retina cells as well as how the adhesion molecules function to bind cells together. Antibodies against cellsurface molecules will be important tools in these investigations.

Research on cell adhesion has mushroomed in recent years and has led to numerous demonstrations of selective cell affinity and to the identification of additional developmentally regulated cell surface molecules that are involved in cell-cell recognition. One commonly encountered cell recognition and adhesion mechanism involves the interaction of cell surface carbohydrate-binding proteins **(lectins)** with complementary carbohydrates on adjacent cell surfaces. Lectins are proteins and glycoproteins that have multiple carbohydrate-binding sites and can therefore link cells together by binding to the carbohydrate moieties of cell surface glycoproteins. Lectins preferentially bind to certain carbohydrates; this specificity can be demonstrated experimentally by using specific sugars, which if they bind the lectin will block lectin-mediated cell-cell association. For example, galactose will block adhesion mediated by a galactose-specific lectin.

The carbohydrate-binding property is also used to purify lectins by passing a mixture of cell surface molecules through an affinity chromatography column containing the appropriate carbohydrate that is bound to support media. The lectin will bind to the carbohydrate and may be eluted from the column by using the appropriate reagents. Lectins are frequently assayed by their ability to agglutinate trypsin-treated erythrocytes, which contain surface carbohydrates that bind the lectins. The carbohydrate specificity of the lectin is examined by determining which sugar will block the agglutination.

Lectin-mediated aggregation has been demonstrated in a number of developing systems. One example is slime mold aggregation. Lectins have also been demonstrated in chick embryonic tissues. Chick embryo lectins are developmentally regulated; that is, they are prevalent at certain stages in the development of particular tissues. Their probable developmental significance is illustrated by the observation that interference with lectin activity inhibits the cohesion of chick embryo cells, induces changes in cell morphology, and promotes cell migratory activity *in vitro* (Milos and Zalik, 1981). Lecti are also found in amphibian embryos. Lectins are not restricted to embryos; in so cases lectin activity associated with particular embryonic cells can demonstrated for different cells in the adult. The lectins apparently play certain roles in development and are utilized of other purposes in the adult.

Lectins have also been used as tools to study cell surface glycoproteins by their selective binding to the carbohydrate moieties of glycoproteins. An example of a commonly used lectin is **concanavalin A,** which is specific for glucose and mannose residues. Lectin-binding properties of cells change during development, which implies that cell surface g1ycoproteins have changed. Modulations of both the glycoprotein and 1 tin compositions of cell surfaces likely play an important role in mediating the changing associations between cells that occur during development.

We have discussed here only a few of many examples of specific cell adhesion mechanisms that have been described in the literature. will be interesting to learn from further investigation whether the mechanisms are unique and unrelated to one another or if they are permutations of a small number of universal mechanisms of cell-cell as association.

Intercellular Junctions

As cells that have been migrating in an embryo adhere to one another and become fixed into a cohesive unit, they form intercellular contaacts that are important to the function of the unit itself. **Gap junctions** between cells are channels of communication through which ions and small molecules may pass. Thus, there is a potential for exchange molecules that carry developmental information through sheets of cells that are interconnected. Whether gap junctions actually play such a role in development remains to be seen. **Tight junctions** are found between cells of an epithelium. These junctions seal the spaces between cells and prevent passage of material across the epithelial sheet. **Desmosomes** are points of adhesive contact between epithelial cells. Inside the cell they serve as points of attachment for a class of intermediate filaments called **tonofilaments,** which (at least in certain cultured cells) are composed of keratin.

Changes in Surface Properties of Malignant Transformed Cells

The formation of a malignant tumor is a case of abnormal development; tumor cells grow without restraint and may dissociate from the tumor and spread to various parts of the body **(metastasize)** to establish secondary tumors. Considerable evidence exists that the properties that characterize normal cells become modified in characteristic ways in malignant cells. Much of this evidence has been obtained from in vitro studies. Cells growing in vitro that are derived from tumors or that are capable of forming tumors in animals are called **transformed cells.** The study of transformed cells has led to an improved understanding of the properties of tumor cells.

As previously discussed, normal cells in tissue culture move along the substrate until they encounter another cell; contact results in inhibition of movement. If normal cells are surrounded on all sides by other cells, movement ceases, and a monolayer is produced in which the cells have a reduced rate of cell division. On the other hand, when migrating transformed cells contact one another, migration may not be inhibited; instead, transformed cells may crawl under one another (underlap) and continue to multiply to form multiple layers of cells on the substrate. This failure of transformed cells to respond to normal controls over migration and growth may explain why cancerous cells can invade other tissues and multiply at an abnormally high rate.

The migratory behavior of normal tissue culture cells is highly dependent upon the adhesive contacts they make with their substrate. Transformed cells are much less adhesive than normal tissue culture cells. As a consequence, transformed cells do not spread out on the substrate when they migrate but remain more compact and rounded with numerous blebs and microvilli adorning their surfaces. This difference is clearly evident when comparing the morphologies of normal cells with cells of the same type that have been transformed with viruses.

If transformed cells are grown on a more adhesive substrate, they revert to a more normal morphology. This observation is consistent with the hypothesis that transformed cells are less adhesive than normal cells, and that this difference in adhesiveness is responsible for the morphological and behavioral differences between normal and transformed migratory cells.

As we have learned, adhesive properties of cells are dependent upon cell surface characteristics. Ther e, we would expect significant differences in cell surface composition of transformed cells as compared to normal cells. Such differences have been well documented. They include changes in glycop in composition and distribution an increased mobility of glycoproteia in the phospholipid bilayer, the appearance of new surface antigens, and the loss of certain cell membrane proteins. One of the most dramatic changes is the reduction in the amount of fibronectin on the surfaces of some transformed cells. Addition of fibronectin to cultures of transformed cells causes these cells to display more normal adhesiveness, contact inhibition of movement, and many aspects of normal morphology. This result confirms that the absence of fibronectin in transformed cells plays an important role in the normal cell behaviour. A more complete understanding of the roles of cell surface molecules such as fibronectin in regulation of cell behavior should assist in our understanding of both normal development and tumorigenesis.

MORPHOGENESIS OF PLANT CELLS: ROLE OF MICROTUBULES

The presence of rigid walls around plant cells places constraints on the cells that prevent plants from constructing new organs by forming new cell associations or changing the shapes of preexisting cells. The formation of plant organs such as roots, stems, and leaves is directly dependent upon mitosis. Preexisting cells in the growing tips of stems and roots divide by formation of new cell walls between the daughter cells at telophase, and the daughter cells and their enveloping cell walls expand. The mode of expansion and the modeling of the walls of cells produced by mitosis determine the final shapes of the cells, and the combined shapes of cells produced at the growing tips of plants determine the shapes of the structures that are formed. Thus, the morphogenesis of plants depends upon control of the plane of cell division, the directions of expansion of cells, and the shapes of the cell walls of daughter cells after division. These processes are all influenced by microtubules.

This process begins with the formation of a **phragmoplast,** a region of vesicles that associate with the microtubules that persist after chromosomal separation. These vesicles, apparently derived from the Golgi apparatus, fuse with one another to form a membranous **cell plate,** within which the primary cell wall forms. The apposing membranes of the cell plate become the plasmalemmae of the daughter cells. The cell wall is formed from microfibrils of cellulose that are laid down on the apposing surfaces of the daughter cells. Permanent gaps remain in the cell plate to become the **plasmodesmata,** through which adjacent cells retain functional continuity.

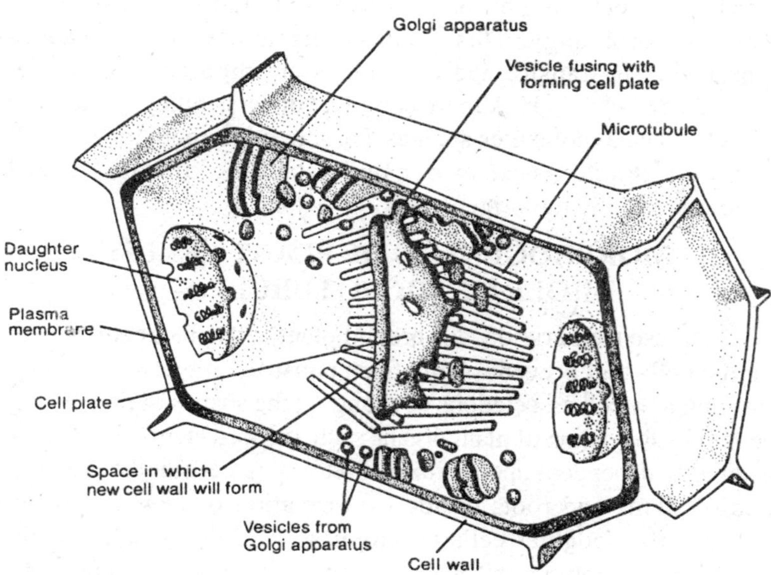

***Figure 6.4** : Schematic representation of a cell of a higher plant as seen at telophase in mitosis. In the phragmoplast region, consisting of membranes and microtubules, a cell plate forms and grows until it separates the cytoplasm into two daughter cells. The cell plate develops as a membrane-delimited structure enclosing a space in which a new cell wall will form. The Golgi apparatus contributes many vesicles to the phragmoplast membrane. The vesicle membranes apparently are incorporated into the membrane of the cell plate, and the vesicle contents enter the forming cell wall.*

Microtubules play key roles in the plant cell cycle (for review, see Gunning and Hardham (1991). Prior to mitosis a circumferential band of microtubules forms in the periphery, or cortex, of the cell, This band, which is called the **preprophase band,** girdles the cell in a plane perpendicular to its long axis. Nearly all microtubules disappear from the remainder of the cell at this time. The site of the next division plane is predicted by this band, since the new cell plate will later bisect' the zone that the microtubule band occupies. When the cell enters mitosis all peripheral microtubules are lost, and microtubules are now present in the spindle. Next, microtubules are found in the phragmoplast, in which the cell plate later develops (at the previously determined site), These microtubules may be responsible for guiding the Golgi vesicles to the division plane. Immediately after cytokinesis very few microtubules are present in the daughter cells. After a short hiatus they appear in the cortex aligned parallel to the plasmalemma. These short, overlapping, parallel microtubules underlie most of the cell surface. They are oriented perpendicular to the direction of cell elongation and form extensive microtubule networks called **interphase** cortical arrays.

The interphase microtubules play a key role in the shaping of cells after division. Cell walls are forced to expand by the turgor pressure that builds inside the daughter cells. When the cell wall is being stretched, additional cellulose fibers are deposited. Their orientation is determined by the interphase cortical microtubules that underlie the cell membrane that is, the cellulose fibrils are laid down parallel to the microtubules. The orientations of these cellulose fibrils, in turn, determine the direction of any future cell wall expansion. Cellulose fibers are hard to stretch -but can be more readily separated from one another. Thus, cells tend to expand at right angles to the orientation of cellulose microfibrils. Since their orientation was previously determined by the orientation of microtubules, the forces that orient the microtubules play a key role in plant cell morphogenesis.

The utilization of common cell constituents such as microtubules for specific purposes in modeling cellular shape indicates that morphogenesis occurs by controlling the function of universal cell machinery rather than by forming a specialized mechanism. We are reminded of E.B. Wilson's statement, that development

> . . . is nothing new. It involves no breach of continuity, and is but a continuation of the vital processes going on in the parental body. What gives development its marvelous character is the rapidity with which it proceeds and the diversity of the results attained in a span so brief.

7

Initiation of Development

The union of the haploid genomes of the egg and sperm at fertilization produces the diploid zygote genome that directs the development of the embryo and specifies the phenotypic characteristics of the resulting adult organism. The impact of the zygotic genome on development is not detected in many animal species, however, until after a considerable delay. Instead, the developmental program is initiated by constituents of the egg cytoplasm. Experimental demonstration of this role of the egg cytoplasm was first provided during the late 1800s and early 1900s by echinoid species hybridization studies. In these experiments, eggs of echinoids were fertilized in the laboratory with sperm from different species (or even different genera), and the characteristics of the hybrid embryos were monitored. In such crosses the morphological characteristics of cleavage-stage embryos (such as the form and rate of cleavage) are found to be entirely maternal, whereas paternal characteristics are not evident until later stages of development. These experiments are interpreted as showing that the cleavage process is directed entirely by the cytoplasm. The embryonic nucleus-which has both maternal and paternal genes-is excluded from participation in the earliest developmental events and does not influence development until somewhat later. The exact timing of the intervention by the embryonic nucleus.

The absence of embryonic nuclear involvement in the regulation of cleavage is dramatically demonstrated by a series of experiments conducted by E.B. Harvey (1976, 1980), who developed a procedure to produce enucleated fragments of sea urchin eggs, which could then be parthenogenically activated to initiate development. The behavior of these fragments (called **merogones)** is entirely attributable to the components of the cytoplasm. In Harvey's procedure, unfertilized eggs are centrifuged in a sucrose solution of the same density as the eggs. The eggs remain suspended during centrifugation, and their contents stratify the eggs elongate and assume a dumbbell shape before the centrifugal forces separate them into two fragments. The light half always contains the nucleus, and the heavy fragment is enucleated. Parthenogenic activation of these enucleated fragments can be achieved by treating the with a solution of hypotonic seawater. These **parthenogenic merogones** go on to cleave and form abnormal blastulae. The enucleated blastulae superficially resemble nucleated blastulae and can even hatch from the fertilization envelope. However, they lack a blastocoel and ha a reduced number of cilia, which are short and irregularly distributed No development beyond the blastula stage is observed.

Cleavage in the absence of a functional nucleus has also been de onstrated in the frog, *Rana pipiens*. Achromosomal embryos are prepard by a two-step operation. First, sperm are irradiated with x-rays to destroy the chromatin; then the irradiated sperm are used to activate.eggs. After activation the haploid egg nuclei are removed manually. These embryos, which contain only functionally inactive chromatin derived from the irradiated sperm nucleus, undergo somewhat abnormal cleavage and form partial blastulae. Histological sections of the partial blastulae confirm that most cells lack chromatin. Irradiated sperm chromatin is found in a small proportion of cells but these cells cleave to the same extent as those without chromatin. Thus, the presence of this chromatin is not correlated with the ability to cleave.

An alternative to removal of the nucleus is the inhibition of nuclear function by using actinomycin D or α-amanitin, both of which interference with RNA synthesis. The use of these inhibitors mimic some aspects of physical enucleation, since the nucleus is prevented from carrying on transcriptional activities. Because these transcriptional in hibitors also have other toxic side effects on cells, interpretation of experiments in which they are used must be made with caution. Although often inconclusive by themselves, experiments employing inhibitors can, however, provide valuable corroborative evidence for data obtained other techniques. Inhibitors were first used to assess the necessity of nuclear function during early sea urchin development. Although treatment of these embryos with actinomycin D severely curtails RNA synthesis, cleavage still occurs, and the embryo proceed as far as the mesenchyme blastula stage before development halted. Similar results have also been obtained with animal from every level of the evolutionary scale with the exception of maimass. A partial list of inhibitor studies on nonmammasian embryos presented in Table 7.1. These inhibitor studies reinforce the concept that initial embryonic development is independent of *de novo* (*i.e.,* new RNA synthesis. However, this generalization cannot be applied to mamalian embryos without qualification. The evidence from similar studies with mammalian embryos indicates a greater reliance on *de no* transcription in that group.

TABLE 7.1 : EFFECTS OF INHIBITORS OF *RNA* SYNTHESIS ON EARLY DEVELOPMENT

Organism	**Observations**	**References**
Coleopteran insects	Arrest at blastoderm stage with either actinomycin D or α-amanitin	Lockshin, 1966; Maisonhaute, 1977
llyanassa obsoleta (marine snail)	Arrest at gastrulation with actinomycin D	Collier, 1966; Feigenbaum and Goldberg, 1965
Sea urchin	Arrest at mesenchyme blastula stage with actinomycin D	Gross and Cousineau, 1964; Gross et al., 1964; Summers, 1970
Ascidia nigra (ascidean)	Arrest at blastula stage with either actinomycin D or α-amanitin	Brachet et at., 1964; Brachet et al., 1972
Rana pipiens	Arrest at blastula stage with actinomycin D	Wallace and Elsdale, 1963
Fundulus heteroclitus (marine teleost)	Arrest at blastula stage with actinomycin D	Wilde and Crawford, 1966

MATERNAL REGULATION OVER EARLY DEVELOPMENT

The ability of the egg cytoplasm to support initial development without embryonic nuclear involvement indicates that preparations for cleavage are made during oogenesis, and that the cleavage mechanism itself must be a product of maternal gene expression rather than of zygotic gene expression. Thus, much of the gene transcription that occurs during oogenesis is in preparation for cleavage, and the effects of this gene activity are seen in the zygote rather than in the oocyte. Genetic evidence for this interpretation has been obtained from a number of genetic mutants with so-called **maternal effects**, which are specific phenotypic traits in the embryo that are determined by the maternal rather than the embryonic genotype. Maternal effect mutants provide some of the most convincing evidence that early development depends to a large extent on substances that preexist in the egg.

The best known example of a maternal effect is that of shell coiling in the snail *Limnaea peregra*. Coiling is usually *dextral* (opening of the shell is to the right). In rare individuals the coiling is reversed, or *sinistral* (opening of the shell is to the left). The direction of coiling is determined by genes at a single locus, with the wild-type dextral allele (+) acting as a dominant to the sinistral allele (s). However, the direction of coiling is determined not by the genotype of the developing snail but by the genotype of its mother. A sinistral (s/s) mother will produce only sinistral offspring, even though the progeny may receive the dominant dextral gene (+) from their father.

Note that the shells of the progeny are not necessarily like the mother's shell but are *determined by her genotype.* For example, a sinistral +/s mother has only dextral progeny, and a dextral s/s mother has only sinistral progeny.

Although shell coiling is a trait seen in adult life, coiling is actually imposed upon the adult by the spiral mode of cleavage, and the direction of coiling depends upon the orientation of the cleavage spindle. Thus, the spindle of sinistral embryos is tipped to the left, whereas that of dextral embryos is tipped to the right. However, the orientation of the spindle is determined *during oogenesis* and is under the control of the maternal genome. Consequently, the maternal genome regulates the direction of coiling in her progeny *regardless of their genotypes.*

A recent experiment suggests that the direction of cleavage in these embryos is regulated by a substance present in egg cytoplasm. Freeman and Lundelius (1982) have reported that cytoplasm from dextral eggs injected into sinistral eggs causes the recipients to cleave in a dextral pattern. The reciprocal cytoplasmic transfer is ineffective, however: Sinistral cytoplasm will not reverse the cleavage pattern of dextral embryos. Therefore, expression of the dextral gene during oogenesis produces a cytoplasmic substance that causes dextral cleavage after fertilization. In the absence of this substance, cleavage will be sinistral.

Injection of dextral cytoplasm is only effective in "curing" the sinistral condition if done before second polar body formation. Thus, the direction of cleavage must be determined at this early stage. This provides a probable explanation for the failure of dextral sperm to influence. the cleavage pattern (and subsequently, the direction of shell coiling) of sinistral eggs: The dextral genes of fertilizing sperm nuclei are not yet capable at this early stage of directing the synthesis of the cytoplasmic product (remember that the pronuclei have not yet fused). Therefore, the direction of cleavage is governed entirely by the preexisting egg cyto plasmic components, which were produced under the control of the maternal genotype. The nature of the dextral substance is unknown, but the microinjection technique provides an ideal assay system to test substances extracted from dextral egg cytoplasm.

A very large number of maternal effect mutants are known in *Drosophila,* providing evidence for the extent of the maternal influence over the embryo. For example, *every gene* that is known to affect the formation of the cellular blastoderm is a maternal effect mutant. Thus, gene control over this vital stage in early development is apparently mediat' through egg constituents that are produced by the transcription of ternal genes during oogenesis.

One of the best studied maternal effect mutants in *Drosophila*

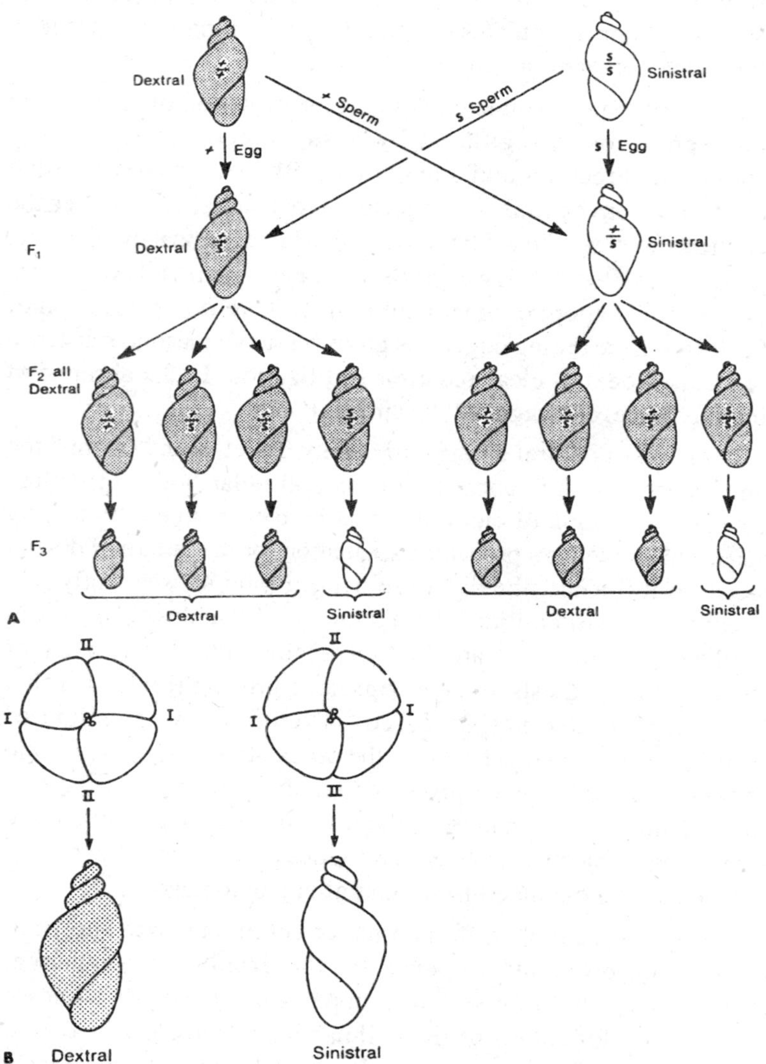

***Figure 7.1 : Shell coiling in the snail,* Limnaea peregra.**
A, Pattern of inheritance.

deep orange (dor), a sex-linked eye color mutant. In addition to its effe on eye color, the gene is also involved in early development. When *+/dor* females mated to *dor/Y* males, their *deep orange* progeny develop normal However, *dor/dor* females do not produce viable *deep orange* progeny Apparently, the wild-type allele of *deep orange* produces a substance that is essential for normal development. This substance is incorporated in all eggs produced by *+/dor* females, and these eggs are capable of nor development *regardless of the genotypes of the sperm that fertilize the* On the other hand, eggs of *dor/dor* females are deficient in this substance. These eggs will not develop if fertilized by dor-bearing sperm. However if the sperm carry the wild-type allele, about half of the + Idor daughter survive. This result is interpreted to mean that expression of the + gen after fertilization may be sufficient to provide the missing substance that is necessary for development.

Further evidence that there is indeed a cytoplasmic substance produced by expression of the wild-type allele of *deep orange* comes from experiments by Garen and Gehring (1972), who injected cytoplasm wild-type eggs into *dor* embryos at the syncytial blastoderm stage. About one third of the injected embryos developed to the stage of embryo movement, a stage never attained by uninjected *deep orange* embryos. As expected, the cytoplasm of *dor* eggs is not effective in repairing the *dor* developmental defect.

Maternal effect mutants have also been discovered among higher organisms, particularly the amphibians. An example is *ova deficient (o)* in the Mexican axolotl, *Ambystoma mexicanum.* Progeny of o/o females die during gastrulation, regardless of the genotypes of the sperm that fertilize the eggs. Cleavage is normal until the late blastula stage, when the cleavage rate becomes reduced. Gastrulation is initiated, but it is abortive. This block can be corrected experimentally by injecting the contents of the germinal vesicle of a normal (+ /o or + / +) ovarian oocyte into c/o eggs following fertilization. The recipients will develop well beyond the stage of the uninjected eggs, and in some cases develop into swimming larvae. Correction of the defect can also be obtained by injection of normal egg cytoplasm after germinal vesicle breakdown, but injection of cytoplasm of ovarian oocytes is only slightly effective. Apparently the normal germinal vesicle contains a factor (the product of the wild-type allele) that accumulates in the germinal vesicle and is released into the cytoplasm at germinal vesicle breakdown. This factor (called o^+ substance) is essential for normal development. Unlike *dor,* the o+ substance is only produced during oogenesis and cannot be supplied by expression of the paternally derived wildtype allele of the embryonic genome. Further characterization of the o+ substance indicates that it is a protein, as shown by its heat lability, its inactivation by the protease trypsin, and its precipitation by ammonium sulfate.

The Oogenic Messenger Hypothesis

The overwhelming evidence that early development is regulated by components of the egg cytoplasm has stimulated a search for the identity of these substances and for their modes of action in mediating the influence of the maternal genome. This search has been a key area of investigation during the past decade and is one of the major success stories in developmental biology. Most of this research has been conducted on sea urchin and amphibian embryos, but the principles derived from work on these organisms are nearly universal.

As with all cells, the functional properties of the fertilized egg and the blastomeres of the cleaving embryo depend upon the proteins they contain. These cells are specialized for rapid mitotic division, utilizing stored energy in the form of yolk. Accordingly, they require structural proteins for the production of mitotic spindles and a vast new membrane system, including a plasmalemma, nuclear envelope, and endoplasmic reticulum for each blastomere. Likewise, they must produce the enzymes for their major metabolic functions, including energy utilization and DNA synthesis. Some proteins are incorporated into the oocyte cytoplasm during oogenesis. However, these components are *not sufficient* to sustain the embryo through the rapid cleavages of initial development without additional protein synthesis. This conclusion is suggested by the results obtained when embryos are treated with puromycin. This inhibitor, which prevents protein synthesis, causes a rapid inhibition of embryonic development. Thus, protein synthesis is apparently of immediate consequence to the embryo in the maintenance of developmental functions. Protein synthesis increases within minutes of fertilization of the sea urchin egg, as a component of the activation response. In amphibians, protein synthesis increases at meiotic maturation in anticipation of fertilization and undergoes an additional increase after sperm entry. The puromycin studies sugg-est that this protein synthesis is *essential* for development to occur.

A parodox becomes immediately apparent. Protein synthesis re-quires messenger RNA, transfer RNA, and ribosomes, all of nuclear origin. And yet we have learned that nuclear function-indeed, the very presence of a nucleus-is *not essential* during early development. We must therefore conclude that the cytoplasm of the egg contains the complete protein synthetic machinery (i.e., messenger RNA, transfer RNA, and ribosomes) in an underutilized form, and activation of development allows utilization of this preexisting store of information. Striking evidence supporting this conclusion comes from experiments in which protein synthesis is demonstrated in the absence of nuclear function. For example, protein synthesis occurs at near normal levels in enucleated eggs of *Rana pipiens* in activated sea urchin merogones, and in sea urchin zygotes treated with actinomycin D. Not only is protein synthesis maintained in the absence of nuclear function, but the *qualitative pattern* of protein synthesis is apparently normal in enucleated *Rana pipiens* eggs. These results suggest that the egg cytoplasm is preprogrammed to direct the protein synthesis that is essential to guide the development of the embryo through cleavage and blastulation.

The hypothesis that messenger RNA is stored in the oocyte cytoplasm for utilization after fertilization has stimulated consid-erable research to demonstrate its existence more directly. Further-more, since these **oogenic messengers** are hypothesized to direct initial embryonic development, investigators have sought to identify specific messengers and determine the roles they play in initial development.

One means of demonstrating the existence of oogenic messengers is to extract RNA from unfertilized eggs and use it to synthesize protein *in vitro.* The *in vitro* synthesized polypeptides can then be analyzed following electrophoresis. In this way not only can the *existence* of a pool of stored mRNA be demonstrated, but also the *identities* of some specific messengers can be inferred by examining the polypeptides they encode.

The *in vitro* translation procedure demonstrates the presence of the most prevalent messenger RNA molecules (i.e., messengers, each of which is present in very large numbers). The remaining messengers (which constitute the vast majority of the RNA complexity) are present in such small numbers that their translational products cannot be detected readily by current technology. The prevalent messengers, by producing a large amount of protein when they are translated, present a technical advantage, since the proteins they encode can be analyzed. Furthermore, the messengers themselves may be analyzed more readily than the less prevalent transcripts.

Infante and Heilmann (1981) analyzed the *in vitro* translation products of polyadenylated RNA isolated from unfertilized sea urchin eggs by two-dimensional gel electrophoresis. The presence of poly (A) tracts on most messenger RNA molecules facilitates their isolation. More than 400 polypeptides are synthesized on the oogenic templates. Some of the spots on this gel have been identified. For example, actin is located in the region labeled A, and tubulin is located in the region labeled T. Consequently, the messengers for these proteins are present in the unfertilized egg.

One of the most prevalent classes of oogenic messengers in sea urchin eggs is the set coding for histone proteins. These messengers are readily separated from the remainder of oogenic messengers by sucrose density gradient centrifugation since the RNP particles containing the histone messengers complexed with protein sediment at 20S, whereas the remaining messengers are in particles that are larger and more variable in size. The presence of histone messengers in these particles is demonstrated by *in vitro* translation experiments.

The *utilization* of oogenic messengers in the first hours after fertilization in sea urchin zygotes is indicated by the observation that RNA extracted from the polysomes of cleavage-stage sea urchin embryos produces exactly the same proteins as the RNP-associated RNA isolated from unfertilized eggs when translated *in vitro*. The presence of these RNAs on polysomes indicates that they are functioning as messengers. As we shall discuss later in this chapter, the vast majority of these messenger molecules are of oogenic origin. The similarity between the *in vitro* protein synthetic patterns suggests that there is no qualitative selection of these messengers in the first few hours after fertilization; each kind of preexisting messenger is translated.

The utilization of oogenic messengers after fertilization is also indicated by the fact that synthesis of proteins known to be encoded by these messengers occurs in the absence of *de novo* RNA synthesis. For example, tubulin is synthesized by sea urchin merogones that have been parthenogenically activated. The synthesis of all five classes of histories occurs in cleavage- and blastula-stage sea urchin embryos in the presence of actinomycin D, which drastically inhibits *de novo* synthesis of RNA.

Evaluation of the oogenic messenger population by *in vitro* translation or by the use of inhibitors is indirect. The use of cloned probes permits direct identification of specific messengers and provides a sensitive means of monitoring for the presence of these transcripts in unfertilized eggs and their utilization after fertilization. Actin and histone transcripts have been identified in this way. The most detailed studies have focused on the histone messengers. The egg of *Strongylocentrotus purpuratus* contains messengers for each of the five classes of histories. Together these messengers constitute approximately 5 to 10% of the total quantity of oogenic messenger RNA. These messengers code for modified histories that are only synthesized during initial development and are assembled into the chromatin of cleavageand (in some cases) blastula-stage embryos. We shall discuss the modified histories of early sea urchin embryos later in this chapter.

The prevalent messengers account for most of the total mass

of mRNA and, as we have seen, present major advantages for analyzing oogenic messenger RNA. However, they account for only a very small minority *of* the total number *of different* messengers that are present (i.e., the messenger complexity), since a very large number of messengers are present in only a few copies each. The latter are collectively called the **complex class of messenger RNA,** as distinguished from the former, which constitute the prevalent class of messenger RNA.

A measure of the complexity of sea urchin oogenic messenger can be obtained by determining the level of hybridization of RNA to total single copy DNA, the class of DNA that contains most structural genes. The complexity of sea urchin egg messenger RNA is approximately 37×10^6 nucleotides, which is equivalent to approximately 20,000 to 30,000 different mRNA sequences. As we shall discuss subsequently, the vast majority of these sequences are being translated at the 16-cell stage. Thus, most of these sequences are indeed oogenic messengers for utilization during development.

Oogenic messengers have been demonstrated in amphibian eggs in much the same way as for sea urchin eggs. Most of the polyadenylated RNA in fully grown *Xenopus* oocytes is in stored RNP particles. The potential utilization of this RNA as messenger is indicated by its ability to be translated *in vitro*. Two-dimensional gel electrophoresis of the proteins synthesized *in vitro* by RNA extracted from unfertilized *Xenopus* eggs reveals an array of proteins, including actin and histories. The messengers for the latter are nonpolyadenylated, whereas historic messengers in oocytes are mostly polyadenylated. Complexity measurements on messenger RNA in *Xenopus* oocytes reveal that they contain approximately 20,000 structural gene transcripts. This estimate is remarkably similar to the estimate obtained for sea urchin egg messenger RNA.

The stored RNA in the unfertilized egg cytoplasm is the informational pool that the embryo utilizes to support the protein synthesis that is necessary for the initial phase of postfertilization development. The amount of stored oogenic mRNA in eggs has been estimated by various investigators. Davidson (1986), who reviewed the literature on this subject, has concluded that oogenic mRNA constitutes approximately 1.5 to 3% of the total RNA of the sea urchin egg and about 1 to 2% of the total RNA in the fully grown *Xenopus* oocyte.

Ribosomal and Transfer RNA for Initial Development

The rapid initiation of protein synthesis that characterizes the beginning of embryonic development requires that the translational machinery be poised for immediate utilization by the oogenic messenger RNA when it is mobilized. Protein synthesis in the absence of RNA synthesis (either in transcriptionally inhibited embryos or in activated enucleated eggs) demonstrates that the cytoplasm of the mature egg contains the complete protein synthetic machinery, including ribosomes and transfer RNA. The vast majority of the RNA in an unfertilized egg is 18S and 28S ribosomal RNA, which are integral components of the cytoplasmic ribosomes. We have previously discussed the mechanisms by which the synthesis of these large amounts of rRNA is facilitated during oogenesis.

The magnitude of the ribosomal store in the egg is beautifully illustrated by the anucleolate mutant of *Xenopus laevis*. Wild-type X. *laevis* have two nucleoli (2-*nu*) in each diploid cell-one per haploid chromosome set. Elsdale et al. (1988) described a heterozygote mutant with one nucleolus per cell (1-nu). The heterozygotes lack a nucleolar organizer on one of two homologous chromosomes. This deficiency is inherited in typical Mendelian fashion. Thus, when two heterozygotes are mated, the progeny fall into three classes having two, one, or no nucleoli per diploid cell in a 1:2:1 ratio, respectively. The progeny lacking nucleoli *(0-nu* or anucleolates) appear perfectly normal as embryos but are retarded after hatching and die as swimming tadpoles before feeding stages.

Since the nucleolar organizer contains the genes that code for ribosomal RNA, the anucleolate mutant embryos lack the ability to synthesize this RNA. In this experiment, anucleolate (0-*nu*) and control (1-*nu* and 2-*nu*) embryos were allowed to incorporate radioactive precursor (labeled with ^{14}C) into RNA. The latter was extracted and fractionated by sucrose density gradient centrifugation. Total RNA (i.e., preexisting and newly synthesized) has been determined by optical density measurements, whereas newly synthesized RNA is indicated by radioactivity determinations. The control embryos have synthesized all the classes of RNA that are resolved by this procedure (28S and 18S rRNA and 4S RNA). In contrast, the anucleolates have synthesized 4S RNA, but the ribosomal RNA is entirely preexisting; no new 18S and 28S RNA has been produced. Since the 0-*nu* progeny are derived from heterozygous mothers, the embryos are able to survive until the swimming tadpole stage on the ribosomes made during oogenesis and stored in the egg cytoplasm.

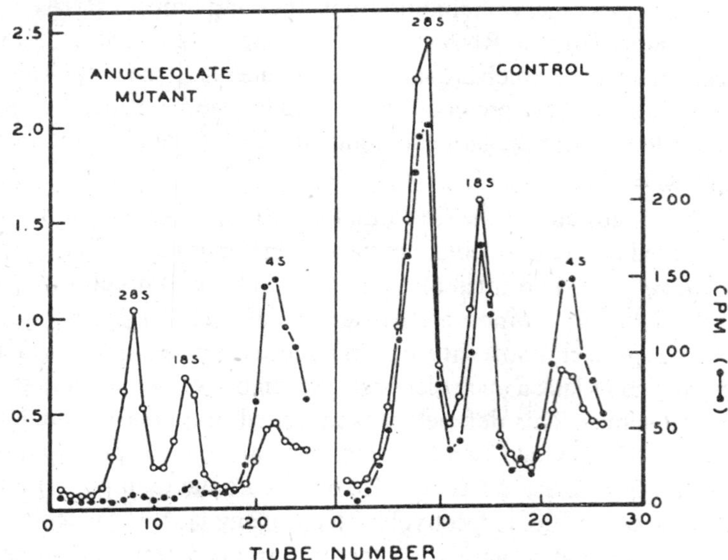

Figure 7.2 : Sucrose density gradient centrifugation of RNA isolated from anucleolate (0-nu) and control embryos. Total RNA is represented by optical density measurements (line with open circles), while RNA synthesized by the embryos is represented by the radioactivity measurements (line with solid circles).

The interesting mechanism that facilitates the accumulation of considerable amounts of 4S and 5S RNA during *Xenopus* oogenesis. Ford (1971) has determined that the late vitellogenic

Xenopus oocyte has more than five 4S RNA molecules and more than seven 5S RNA molecules for each 18S and 28S RNA molecule. The retention of 4S and 5S RNA after oogenesis has been shown by labeling them during oogenesis and recovering the labeled molecules in eggs.

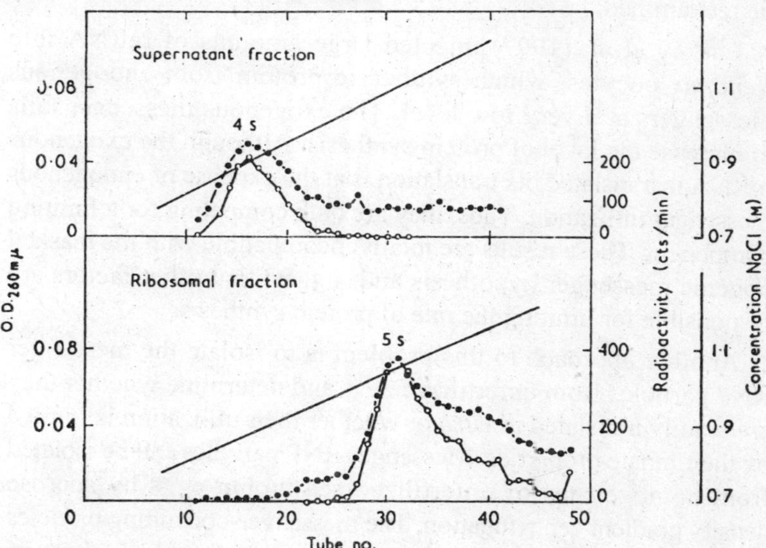

Figure 7.3 : Column chromatography of 4S and 5S RNA from mature Xenopus eggs. Eggs were obtained from a female who had been injected with ^{3}H-uridine. RNA synthesized during oogenesis is radioactive. RNA was isolated from egg supernatant or ribosomal fraction, and low molecular weight RNA from these fractions was chromatographed on protamine-Celite columns. The RNA was eluted from the columns by a NaCl gradient. Molarity of NaCl: solid line. Optical density: line with open circles. Radioactivity: line with solid circles.

Why Are Stored Messengers Inactive in Protein Synthesis?

The low level of protein synthesis before fertilization (in the sea urchin) or maturation (in the amphibian) must result from limitations on the components of the protein-synthesizing system. It has been proposed that the messenger synthesized during oogenesis is stored in the egg cytoplasm in an inactive form and that after fertilization it is mobilized for translation. This is the **masked oogenic** messenger hypothesis. According-to this hypothesis, the messenger is masked in the sense that it is in a nontranslated form and protected from degradation by nucleases, making it extremely long-lived. Although the masked oogenic messenger concept has dominated thought about limitations on protein synthesis in eggs for some time, recent evidence suggests that this concept must be reexamined.

Laskey et al. (1997) injected large amounts of mRNA into *Xenopus* oocytes, which synthesize protein from endogenous messengers at a very low level. The exogenous messenger fails to increase the level of protein synthesis. Although the exogenous mRNA is translated, its translation is at the expense of endogenous messenger utilization. Thus, they are both competing for a limiting component. These results are totally incompatible with the masked oogenic messenger hypothesis and suggest that other factors are responsible for limiting the rate of protein synthesis.

Another approach to this problem is to isolate the messenger RNP particles from unfertilized eggs and determine whether they are readily translated in *vitro* or whether their utilization is limited by their innate properties. Messenger RNP particles can be isolated from homogenates of unfertilized sea urchin eggs by sucrose density gradient centrifugation. The messenger-containing particles sediment more slowly than ribosomes and are found in a fraction that has been called the **subribosomal fraction.** Jenkins et al. (1998) have reported that when these particles are isolated under conditions that preserve their stability, the RNA remains nontranslatable. However, under conditions that do not preserve RNP stability, the RNA can be readily translated *in vitro.* These results suggest that an alteration in the RNP particles after fertilization may allow translation of stored messengers in vivo. More recently, Moon et al. (1982) have reported that the RNA in egg RNP particles is *translatable* in vitro. They conclude that. the RNA is not masked. The results of in vitro translation experiments are heavily dependent upon the techniques used in RNP isolation and translation. Thus, this experimental approach may be inappropriate to determine whether oogenic messengers are "masked."

In the absence of definitive evidence in favor of masking-at least in sea urchins and amphibians-we should consider other alternatives as explanations for the inactivity of stored messengers. One mechanism for limiting access to the translational machinery could be physical sequestration of messengers that would restrict their access to the translational machinery. Perhaps stored messengers are localized to cytoplasmic structures such as annulate lamellae. An unusual form of sequestration has been discovered for a subset of the oogenic histone messengers in the sea urchin egg; they are sequestered in the female pronucleus. Another possibility involves the cytoskeleton. Moon et al. (1983) have demonstrated that actively translated messengers in sea urchin embryos are associated with the cytoskeleton, whereas most messengers in eggs are not associated with it. Perhaps the translation of messengers in the egg is limited by a mechanism that restricts access of messengers to the cytoskeleton.

Finally, it is possible that egg messengers are fully translatable with unrestricted access to ribosomes, but the ribosomes are unable to translate them, perhaps because initiation factors are limiting. We should, however, keep in mind that a variety of mechanisms to limit egg protein synthesis may operate in different organisms and even for different kinds of messenger RNA in the same organism. Investigation of the mechanisms responsible for restricting the utilization of stored messengers is an active, exciting area of contemporary research. The literature in this area should yield much valuable information in the near future.

Embryonic Processing of Oogenic Messenger

After fertilization, the translation of oogenic messengers contributes to an acceleration in the rate of protein synthesis. The mobilization of stored oogenic mRNA and ribosomes to function in protein synthesis is called **recruitment.** As a consequence of this mobilization, ribosomes bind to messenger RNA molecules to form polysomes. Thus, the messengers shift from the subribosomal fraction to the ribosomal fraction. This shift has been detected in sea urchin eggs by tracing the subcellular location of polyadenylated RNA before and after fertilization. Polyadenylated RNA shifts from the subribosomal to the ribosomal fraction within 30 minutes after fertilization.

The active translation of oogenic1 messengers commences with recruitment and the formation of polysome complexes. Humphreys (1991) has demonstrated that most of the RNA being translated on polysomes during the early cleavage stages of sea urchins is of oogenic origin. The increase in polysome content in sea urchin embryos after fertilization is reflected in the increase in protein synthesis during the same period. In *Xenopus laevis* the initial increase in polysome content begins at maturation, and it rises even more after fertilization.

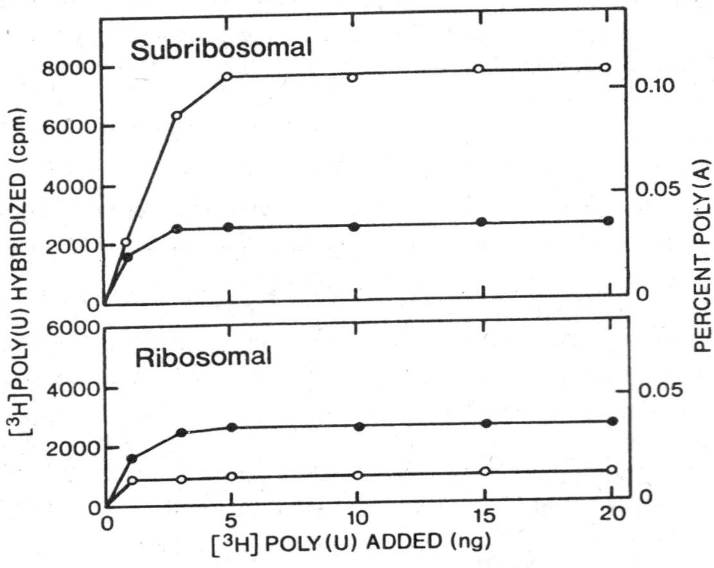

Figure 7.4 : Saturation hybridization between ^{3}H-poly (U) and either subribosomal- or ribosomal-associated RNA from sea urchin eggs before fertilization (line with open circles) and 30 minutes after fertilization (line with solid circles).

In sea urchins the increase in protein synthesis is an integral part of the release of metabolic inhibition after fertilization. The release of inhibition is thought to be triggered by the increase in intracellular pH that follows fertilization. The dependency of the rate of protein synthesis on pH is illustrated by experiments in which manipulation of intracellular pH of eggs is shown to affect the rate of protein synthesis. Sea urchin eggs can be activated with NH_4C1, which generates ammonia that enters the eggs and causes

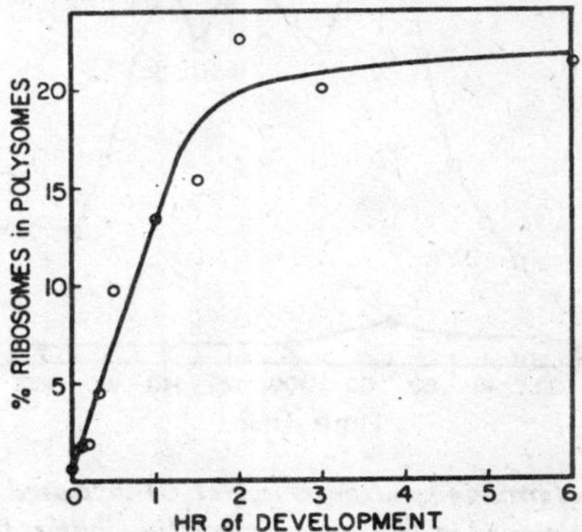

Figure 7.5 : The increase in ribosomes in ribosomes in polysomes after fertilization of the sea urchin egg. The proportion of ribosomes was measured at various times after fertilization.

intracellular pH to rise. The NH_4C1 treatment causes an increase in the rate of protein synthesis. Transfer of eggs to seawater lacking NH_4CI (arrow) causes a drop in intracellular pH and a concomitant fall in the rate of protein synthesis. In this experiment sea urchin eggs of *Lytechinus pictus* were fertilized, and the internal pH was regulated by (1) adding seawater containing sodium acetate, which enters the eggs and lowers intracellular pH, followed by (2) removal of the acetate stimulus, which allows pH to increase once again. The effects of sodium acetate on intracellular pH are shown in Figure; within 25 minutes of adding seawater containing acetate the intracellular pH drops to the unfertilized egg level. These pH changes have profound effects on the rate of protein synthesis. Addition of sodium acetate stops the rise in the rate of protein synthesis, and after the intracellular pH has returned to the unfertilized egg level, the rate of protein synthesis drops off linearly until it, too, reaches the unfertilized egg rate. This effect of reduced pH on protein synthesis is reversible, since resuspending the eggs in seawater lacking sodium acetate allows protein synthesis to increase once again. Clearly, the activation and maintenance of a high rate of protein synthesis are dependent upon the maintenance of an alkaline pH in the egg cytoplasm.

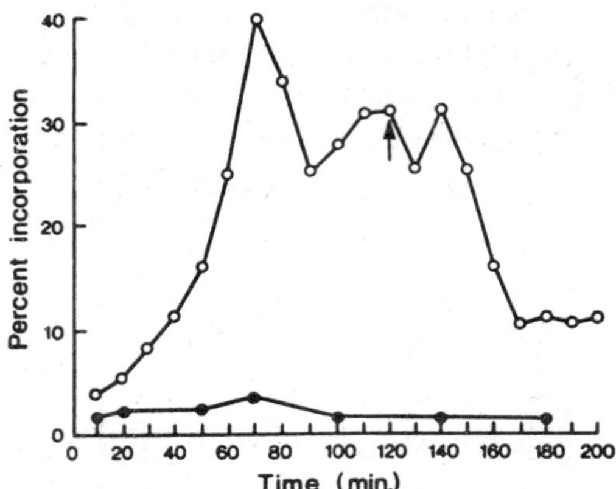

Figure 7.6 : The rate of protein synJthesis in NH_4Cl-activated sea urchin eggs. Unfertilized eggs were placed in either sea water (line with solid circles) or sea water containing NH_4Cl (line with open circles), and the rate of protein synthesis was determined at 10-minute intervals. The rate of protein synthesis is indicated by percent incorporation of ^3H-valine. This was calculated by dividing the amount of isotope incorporated into macromolecules (which are precipitated by trichioroacetic acid) by the total amount of isotope accumulated in the eggs. At 120 minutes (arrow) the eggs were removed from NH_4Cl and resuspended in sea water. Note the subsequent drop in the rate of protein synthesis.

Although the elevation of intracellular pH with NH_4Cl mimics the effects of fertilization on protein synthesis, this response is not the same as the response to sperm. Grainger et al. (1979) observed that the rate of protein synthesis in ammonia-activated eggs lags behind that of fertilized eggs. This lag can be completely overcome if the ionophore A23187 is used to trigger a release of intracellular Ca^{++} in NH_4C1-activated eggs. This result suggests that both the Ca^{++} release and the increase in intracellular pH are necessary for maximal stimulation of protein synthesis.

The acceleration of protein synthesis involves at least two changes in the protein synthetic apparatus. These are the mobilization of messengers and changes in the utilization of messengers by the ribosomes. The mobilization of messengers in sea urchin zygotes is apparently a direct consequence of the pH increase. How this effect is mediated remains uncertain. The relatively low pH that exists before fertilization may maintain the messenger

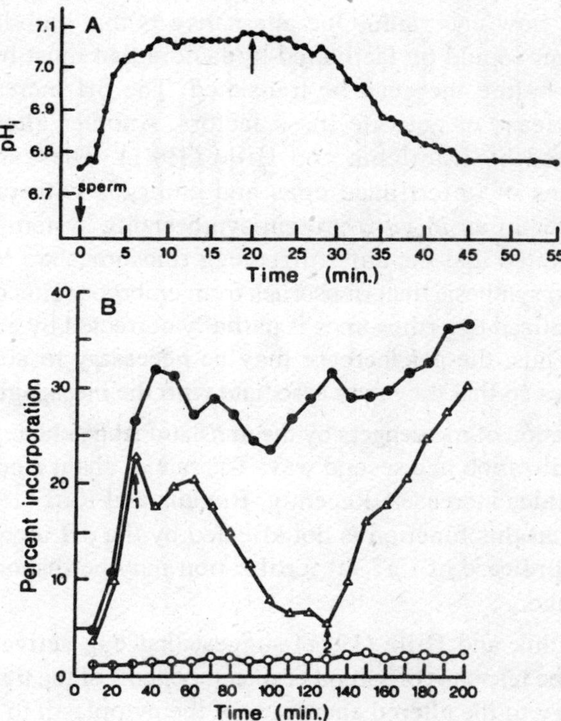

Figure 7.7 : The relationship between intracellular pH and the rate of protein synthesis in fertilized sea urchin eggs. A, Eggs were fertilized and maintained in sea water. Internal pH (pH;) was measured with a microelectrode. At 20 minutes after fertilization (arrow) eggs were exposed to sodium acetate. Internal pH declined, dropping to the unfertilized egg level within 25 minutes. B, The effects of sodium acetate on the rate of protein synthesis (line with open triangles). At 30 minutes after fertilization (arrow no. 1) eggs were resuspended in sea water containing sodium acetate. The rate of protein synthesis dropped to the unfertilized egg level. At 130 minutes (arrow no. 2) the eggs were returned to sea water lacking sodium acetate. Protein synthesis accelerated. For comparison, the rates of protein synthesis for unfertilized eggs (line with open circles) and fertilized eggs maintained in sea water (line with solid circles) are shown.

RNP particles in an untranslatable state, and the pH increase may release the inhibition. Their behavior may be related to a number of specialized proteins associated with the RNA, which are different from those found in RNPs from late postfertilization embryos.

However, as we have previously discussed, masking of RNA in RNPs is now uncertain. One alternative is that mobilization of messengers could be facilitated by factors that must bind to the mRNPs before they can be translated. The pH increase could either release or activate these factors. Another alternative is implicated by Danilchik and Hille (1981). They challenged ribosomes of unfertilized eggs and embryos with exogenous messenger in an *in vitro* protein-synthesizing system and have demonstrated that the unfertilized egg ribosomes are less active in protein synthesis than ribosomes from embryos. This deficiency in unfertilized egg ribosomes is partially corrected by an increase in pH. Thus, the pH increase may be necessary to activate the ribosomes so that they may associate with the messengers.

Utilization of messengers by the translational machinery changes after fertilization in a second way: The rate of chain elongation of polypeptides increases. Recently, Brandis and Raff (1979) have shown that this function is not affected by the pH increase. The temporal release of Ca^{++} at fertilization may be responsible for this change.

Danilchik and Hille (1981) suggest that egg activation may involve the response of a number of components of the translational machinery to the altered chemistry of the cytoplasm to bring the rate of protein synthesis up to the level required to initiate development. Recently, Winkler and Steinhardt (1981) have developed a cell-free protein synthetic system from sea urchin eggs that may provide the assay system that will lead to a resolution of the problem of control of protein synthesis after fertilization.

An important consideration about oogenic messengers concerns possible regulation over their utilization: Are they mobilized en masse, or is there regulation over when individual transcripts will he utilized or, indeed, if they will be utilized at all? .If there were no such regulation, the entire oogenic RNA population would he represented on polysomes throughout initial development. The utilization of prevalent messengers can be inferred by comparing the *in vitro* translation products of polysomal RNA with the products of transcripts stored as RNP. As we discussed earlier,

RNA extracted from the polysomes of cleavage-stage sea urchin embryos produces the same kinds of proteins as the RNA isolated from unfertilized eggs when translated *in vitro*. However, careful comparison of gel autoradiographs of proteins synthesized in *vitro* from polysomal and free messengers from cleavage-stage embryos reveals certain quantitative differences. Thus, there is a translational control mechanism that regulates the number of certain transcripts that are mobilized.

Histone messengers in the sea urchin zygote are also subject to translational control. This is demonstrated by the delayed recruitment of a subset of these messengers as compared to the bulk of oogenic messenger-RNA. Delayed recruitment of histone messengers has been attributed to their unusual mode of storage in the egg: They are stored in the egg pronucleus, rather than in the cytoplasm. Meiosis is completed before the sea urchin oocyte has ovulated. Thus, the unfertilized egg contains a haploid pronucleus that fuses with the male pronucleus after fertilization. In preparation for the first cleavage division the envelope surrounding the fusion nucleus breaks down before metaphase. At this time the sequestered histone messengers are released from the nucleus and can be translated.

The utilization of the complex class of messengers is demonstrated by RNA-DNA hybridization technology. A specific probe for oogenic messengers is made by hybridizing egg RNA with single copy DNA and recovering the DNA from the hybrids. This DNA is then used to detect homologous sequences after fertilization. This DNA has been designated oDNA (i.e., oocyte DNA). The percent of hybridization between the oDNA and embryonic RNA populations is a measure of their relatedness to the original oogenic messenger set. RNA extracted from polysomes at the 16-cell stage is, by definition, the RNA that is being actively translated at that stage, and as shown by Humphreys (1971), it is predominantly of oogenic origin. The oDNA hybridizes with this polysomal RNA to 73% of the reaction with oogenic RNA. In other words, approximately 73% of the oogenic RNA sequences are being translated at this time. Interestingly, oDNA hybridizes with *total* cytoplasmic 16-cell-stage RNA to 100% of the reaction with oogenic RNA, an indication that virtually all of the oogenic sequences are present in the cytoplasm, although they are not all being concurrently translated. One possible interpretation of these experiments is that a portion of the complex class RNA is not functional as messengers.

The possibility that oogenic RNA contains transcripts that are not translated is also. suggested by the observation that most of the total mass of the cytoplasmic complex class transcripts of sea urchin eggs contains repetitive sequences that are interspersed (i.e., covalently linked) with the single copy sequences (Costantini et al., 1980). This suggests that adjoining single copy and repetitive sequences of DNA are transcribed during oogenesis, -and the repetitive sequences are not removed from the transcripts before they are transported out of the oocyte nucleus. Perhaps the incompletely processed transcripts are not messengers but perform some as yet undiscovered function during development. Alternatively, they may be translatable if they undergo processing in the cytoplasm to remove the interspersed regions. However, no cytoplasmic mechanism of this kind has been described. It is interesting and perhaps significant that a major fraction of *Xenopus* oocyte polyadenylated RNA also has this same interspersed sequence configuration. The presence of these transcripts in these two evolutionarily distant groups suggests that this may be a common feature of oogenic RNA with an important function in initial development.

Localization Of Oogenic Messenger

Since oogenic messenger RNA plays a critical role in directing initial development, it is logical to inquire whether it is equally distributed in the cytoplasm *or* if there is a pattern of distribution that may play a role in directing the differential development of regions of the embryo. The most logical embryos in which to search for mRNA distributional patterns are those that contain recognizable cytoplasmic localizations of morphogenic determinants. A coincidence of messenger localization with regions containing morphogenic determinants would suggest a role for the messenger in directing regional development. Ascidian eggs, with their three identifiable ooplasms, are excellent material in which to search for messenger localization. Recent experiments with these embryos do reveal a coincidence of messenger RNA distribution with certain ooplasms.

Jeffery and Capco (1978) used ^{3}H-poly (U) to localize polyadenylated RNA in sectioned *Styela* embryos by in *situ* hybridization. Their results indicate that polyadenylated messenger RNA is enriched in the ectoplasm, whereas the endoplasm has an intermediate concentration of this RNA, and the myoplasm has very low concentrations. Their experiments also show that the characteristic concentrations of poly (A) RNA are retained while the ooplasms are shifted during ooplasmic segregation. This observation suggests that the RNA molecules have an affinity for specific ooplasmic constituents. Recent experiments indicate that the mRNA is hound to the cytoskeleton.

The distribution of *specific* mRNA sequences has also been studied *by in situ* hybridization with cloned probes, which were labeled with I2-I. Two probes were used in this study: a histone gene probe and an actin gene probe. Results with the histone probe indicate that histone oogenic messengers are evenly distributed in the egg and embryo. However, actin mRNA distribution is distinctly asymmetrical. The highest concentrations of actin mRNA molecules are detected in the ectoplasm and myoplasm, with very low concentrations in the endoplasm. The high concentrations of actin messenger in the myoplasm are particularly striking, since very low concentrations of messenger (i.e., poly $(A)^+$ RNA) are found there. Hence, actin messengers must account for a substantial percentage of the messenger RNA in the myoplasm. The significance of the distribution pattern of actin mRNA is uncertain. However, since the myoplasm contains muscle determinants, it is tempting to speculate that the presence of the messenger RNA for a muscle protein is related to muscle differentiation. Further experimentation is necessary to establish such a relationship.

EXPRESSION OF THE EMBRYONIC GENOME

The results of experiments involving species hybridization and enucleation indicate that the initial stages of embryonic development are extensively influenced by the maternal genome, whereas the newly formed embryonic genome seems to have little impact on development at this time. As development proceeds, the oogenic influence wanes, and the embryonic nuclei assume preeminence. The assumption of embryonic nuclear control actually involves two stages. The first is the *synthesis of RNA* by the nuclei of embryonic cells, and the second is the *utilization of the newly synthesized RNA* to produce functional proteins that impart on the embryonic cells a new identity that reflects the new combination of paternal and maternal genes.

Onset of Embryonic Nuclear Control

The concept of delayed nuclear influence over development originated with the echinoid species hybridization experiments conducted at the turn of the century. These experiments pinpointed the mesenchyme blastula stage as the time of onset of embryonic nuclear influence. Before this stage, phenotypic characteristics are entirely maternal; afterward paternal or hybrid characteristics are detected. This shift of responsibilities is illustrated by the behavior of the primary mesenchyme cells. These cells detach from the blastoderm during the late blastula stage and populate the blastocoele. Later, during midgastrulation, they begin to form triradiate spicules and eventually produce the branched skeleton of the pluteus larva. The formation of the primary mesenchyme cells appears to be a responsibility of the maternal genome, whereas the differentiation of these cells to form the skeleton is regulated by the embryonic genome.

A cross between echinoids that differ in the origin and subsequent differentiation of the mesenchyme illustrates this point. *Cidaris tribuloides is* a primitive echinoid that differs from the more advanced sea urchins in certain aspects of primary mesenchyme formation. In *Cidaris* the mesenchyme forms from the tip of the archenteron *after its invagination.* Conversely, in *Lytechinus variegatus,* a typical urchin, the primary mesenchyme is formed *before invagination* of the archenteron, and the cells originate from the base and sides of the presumptive archenteron. When *Cidaris* females are mated with *Lytechinus* males, the characteristics of the cleavage- and blastula-stage embryos are entirely maternal; the primary mesenchyme cells originate after gastrulation has begun, albeit somewhat earlier than in *Cidaris* embryos. However, the site of origin of these cells is at the base and sides of the archenteron-a paternal characteristic. Thus, the timing of their formation is primarily maternally determined, but the mode of their formation is determined by embryonic genes. The results of other hybridization experiments indicate that the skeletal elements of the pluteus, which are formed from the primary mesenchyme cells, assume hybrid characteristics, an indication of embryonic gene expression during their differentiation. Treatment of sea urchin zygotes with actinomycin D confirms the conclusions of these experiments: Primary mesenchyme is formed in the presence of the inhibitor, but the mesenchyme cells do not differentiate to form skeletal elements.

A more precise means of detecting the assumption of control of the embryonic genome over development is to monitor the participation of the embryonic genome in RNA and protein synthesis. We shall now examine the initiation of transcription in sea urchin development and the appearance of these newly synthesized transcripts on polysomes. Where possible, we shall analyze the synthesis of *specific transcripts* and their utilization in protein synthesis.

As we shall see, transcription of the embryonic genome begins during early cleavage. However, as we learned earlier in this chapter, the majority of the messenger translated by cleavage-stage embryos is of oogenic origin. Thus, the expression of embryonic genes can make little impact on protein synthesis during cleavage. One reason for this is that the small number of nuclei during early cleavage is able to produce only a ,small amount of RNA in comparison to the large amount of oogenic messenger stored in the cytoplasm.

The impact of embryonic gene expression increases as oogenic messengers are replaced by embryonic transcripts. By the blastula stage the polysomes contain predominantly *newly synthesized* messengers. This replacement is dependent upon the degradation of preexisting messengers and the accumulation of newly synthesized messengers. This does not happen all at once but is phased in over a period of time and varies, depending on the particular RNA.

One way to monitor the accumulation of newly synthesized messengers is to measure the level of messenger abundance as development proceeds. An increase in abundance of a particular messenger indicates *de novo* transcription of its gene. Changes in messenger abundance during initial development can be determined by hybridization analysis between the RNA and cloned DNA probes. Actin messenger abundance has been determined in this way. The procedure used in these experiments is a modification of the Southern blotting technique. In this procedure (called **RNA blot analysis** or **Northern blotting)** the RNA is transferred by blotting to a filter, on which it is hybridized to labeled single-strand DNA. The results of RNA blot analysis with RNA from *Strongylocentrotus purpuratus* embryos show that two size classes of actin mRNA are present at relatively low levels in the egg and remain at low levels until the blastula stage, when they have increased substantially due to *de novo* synthesis. An interesting variation on this pattern is seen in the related species, *Strongylocentrotus drobachiensis,* in which the actin messengers are more abundant in eggs than in 16-cell embryos and then begin to accumulate in the blastula and gastrula stages. The oogenic actin messenger is apparently degraded shortly after fertilization and is replaced by embryonic messengers at the later stages. Clearly, there are species-specific variations in the longevity of oogenic messengers.

Histone messengers present a unique opportunity to monitor the utilization of oogenic messengers and the expression of embryonic genes, since specialized histone gene families are expressed at specific times during early development. Each family contains genes for the five types of histones. Genes in one family differ in nucleotide sequence from the homologous genes in the other families. Messengers produced by one family are distinguishable from messengers transcribed from another family by their distinct electrophoretic mobility, the specificity of hybridization to cloned histone genes, and the thermal stability of RNADNA hybrids. The latter property is used to distinguish messengers that incompletely hybridize with a DNA probe, since hybrids of this sort have reduced stability. For example, messengers transcribed from the late histone gene family (see following discussion) form hybrids having reduced stability with genes of the early histone gene family.

In some cases the differences in nucleotide sequence among families are sufficient to produce electrophoretically distinct histone proteins. This is most pronounced for the H1, H2A, and H2B histones. The messengers for these histones from different families can be identified by the specific histones they encode in *vitro.*

During very early cleavage stages the histones that are synthesized are called the **cleavage-stage (CS) histone subtypes.** The messengers for CS histones are present in the oogenic messenger population and are also synthesized for a short time after fertilization until the genes encoding them are silenced.

By the eight-cell stage, synthesis of the so-called **early histone subtypes** (sometimes called **a variants)** is initiated. These histones continue to be made until the blastula stage. Like the CS histone messengers, the messengers for the early histones are also present in the oogenic messenger population. However, these two types of messengers are stored in separate compartments in the egg. The CS histone messengers are stored in the cytoplasm along with the messengers for other proteins, whereas the early histone messengers are stored in the female pronucleus. We have discussed previously this unusual mode of messenger storage and its implications for messenger utilization. Synthesis and accumulation in the embryo of transcripts of the early histone genes begin in early cleavage but persist longer than for the CS histone messengers. The messengers transcribed from the early genes reach a maximum level of accumulation by the 200-cell stage, after which they rapidly decline in amount.

A third category of histones, called the **late histone subtypes,** is synthesized from the 16-cell stage through the pluteus stage. By the midgastrula stage these are the only histoncs being synthesized. Transcription of the genes encoding the late embryonic histones occurs as the levels of the early histone messengers begin to decline. As a consequence, late histone messenger accumulation coincides with the disappearance of the early histone messengers. The transition to late histone messenger accumulation is concomitant with the reduction in the rate of cell division that follows cleavage.

The stage-specific synthesis of histones provides an excellent system to study the utilization of oogenic transcripts, initiation of transcription, and the regulation of messenger accumulation and utilization during early embryonic development.

The transition from utilization of oogenic messengers to embryonic messengers shifts the responsibility for protein synthesis to the embryonic genome. As we have seen, this can result in qualitative changes in the historic messenger population. Is there a generalized qualitative change in messengers as a consequence of this transition, or does the repertoire of the messenger population remain much the same in spite of the shift in the source of the messengers? A variety of studies have revealed that a great many sequences utilized in embryonic protein synthesis are represented in both the oogenic and embryonic messenger RNA, whereas others are supplied entirely as oogenic messengers, and still others are supplied entirely by transcription after fertilization (for additional discussion.

As development continues past the blastula stage, the messenger RNA complexity is reduced, reflecting a decrease in the number of different mRNA sequences. This is illustrated by preparation of cDNA from the prevalent polyadenylated RNA isolated from polysomes of blastula-stage sea urchin embryos and the use of this cDNA to detect homologous sequences on gastrula polysomes. Of the prevalent blastula messengers, only about 16% remain prevalent on gastrula polysomes, another 16% are present at reduced levels, whereas the remaining two thirds are absent. However, when *nuclear* RNA from gastrulae is hybridized with the cDNA, it hybridizes to the same extent as blastula mRNA. Thus, all of the abundant blastula messenger species are present in the gastrula nucleus, despite the deficiency of two thirds of them from the polysomes. Clearly, the nuclei of gastrula cells continue to synthesize the abundant blastula messenger sequences, but posttranscriptional mechanisms regulate their utilization as messengers in the cytoplasm. Similar results have been reported for complex-class mRNA sequences, indicating that the reduction of complexity of both prevalent and complex-class messengers is due to posttranscriptional mechanisms.

In addition to synthesizing messenger RNA, the nuclei of sea urchin embryos are active in the synthesis of ribosomal RNA, thus supplementing the large amount of this RNA synthesized during oogenesis and stored in the egg ribosomes. However, the rate of synthesis of rRNA is quite low in comparison to the synthetic rate during oogenesis. Electron micrographs of sea urchin embryo chromatin prepared by the Miller technique (chromatin prepared by this procedure retains RNA in the process of synthesis) reveal that the small number of ribosomal RNA genes that are being transcribed are highly active. The difference in synthetic rates in oocytes and embryos is probably due to a difference in the number of rRNA genes that are synthetically active. The vast majority of rRNA genes remain inactive during embryogenesis. Whether the genes that are active during embryonic development are a small subset of those genes that are active during oogenesis, or whether there is a distinct large set of these genes that are active during oogenesis and a much smaller set that is active after fertilization is not known. The rate of rRNA synthesis is increased at the pluteus larval stage after feeding has begun.

For comparative purposes we shall now examine the onset of embryonic nuclear control over development in amphibians. Most of the research on the molecular biology of early amphibian development has been conducted on *Xenopus laevis*. Hence, this discussion is based upon data obtained with this species.

Transcription is demonstrated in early cleavage-stage *Xenopus* embryos by incorporation of radioactive precursors into RNA. The type of RNA synthesized by these embryos is determined by sucrose density gradient centrifugation and base composition analysis. This RNA is heterogeneous in size, has DNA-like base composition, and is primarily nuclear. Thus, it is primarily HnRNA (heterogeneous nuclear RNA). Not all of the newly synthesized RNA is restricted to the nucleus, however, since it has also been detected on polysomes. Some of it must therefore be messenger RNA. Little mRNA is synthesized during amphibian cleavage, and this RNA decays very rapidly. The minute amounts of newly synthesized messenger RNA that accumulate have little impact on the RNA content of these embryos because the preexisting oogenic RNA is present in overwhelming amounts. In the blastula, messenger RNA accumulation increases markedly because of an apparent stimulation of transcriptional activity.

As we discussed previously, the decay of oogenic messengers and accumulation of newly synthesized mRNA shifts control over development to products of embryonic gene transcription. Hybridization analyses with eDNA probes representative of individual oogenic polyadenylated mRNA sequences have monitored changes in these sequences during early development. Most (80%) of these sequences increase in amount during early development and are therefore produced by embryonic transcription as well as by oocyte transcription. One sequence largely disappears at the late blastula or early gastrula stage. Embryonic synthesis of this sequence is thus of little or no consequence. This study has examined a small handful of literally thousands of oogenic sequences. It is, consequently, premature to generalize on the findings. However, the results suggest that the vast majority of oogenic transcripts are resynthesized during development.

In spite of the resynthesis of oogenic messenger sequences by embryonic gene transcription, *Xenopus* embryos also acquire messenger sequences that do not preexist as oogenic messenger. Schafer et al. (1982) have monitored for the appearance and accumulation of such sequences during development, using cDNA clones prepared from polyadenylated RNAs isolated from *Xenopus* somatic cells. Clones that do not hybridize detectably to oocyte RNA were isolated. RNA blot analyses reveal that messengers that are complementary to these probes appear at various times *after gastrulation*. Presumably, the appearance of each sequence is strategically timed to coincide with a particular developmental event that requires the messenger. More data of this type should provide a detailed chronology of the first appearance of a large number of specific messengers during development and provide a reconstruction of the process of assumption of control over development by the embryonic genome.

We shall now consider the *de novo* transcription of the other major classes of RNA; that is, tRNA, 5S RNA, and 18S and 28S RNA. Unlike messenger RNA, no new synthesis of these types of transcripts has been detected during cleavage. Their synthesis is first detected at the blastula stage, coincident with the acceleration in messenger RNA synthesis. Transcription of both oocyte and somatic 5S RNA genes occurs during oogenesis; this transcription is terminated after oocyte maturation. However, in the embryo only the somatic 5S RNA genes are transcribed, whereas the oocyte 5S RNA genes remain inactive. The differential activation of 5S RNA genes is an excellent system for the analysis of the regulation of gene transcription.

The activation of transcription of tRNA, 5S RNA, and 18S and 28S RNA genes and the acceleration in synthesis of messenger RNA during the blastula stage of *Xenopus* development indicate that the nuclei have undergone a major transition at this stage that results in transcriptional stimulation. In addition to the stimulation of transcription, a number of other parameters change at this time: The rate of cleavage slows, the synchrony of cell division is lost, and cells become motile for the first time. This constellation of events marks the midblastula transition. Newport and Kirschner have demonstrated that the midblastula transition is not due to an endogenous counting mechanism based on either elapsed time, the number of rounds of DNA synthesis, or the number of cleavages. Instead, they proposed that the transition occurs when the ratio of nuclear and cytoplasmic volumes has reached a critical level.

As the number of nuclei increases during cleavage, the total cytoplasmic volume remains constant. Therefore, the nucleocytoplasmic volume ratio increases. Newport and Kirschner (1982b) have proposed that a substance present in excess in the egg cytoplasm (and not synthesized after fertilization) is removed from the cytoplasm by the nuclei. This substance is proposed to maintain the high rate of mitosis found during cleavage. After the nuclei have reached a critical number the substance is reduced to a threshold level, below which the high cleavage rate can no longer be maintained; the slower division rate that ensues allows for the expression of previously suppressed functions such as transcription. The nuclear component may be DNA, which would bind the cytoplasmic substance in stoichiometric amounts (i.e., binding would be directly proportional to the total amount of DNA).

To test these hypotheses, Newport and Kirschner injected plasmids containing a yeast tRNA gene into fertilized eggs, which were prevented from cleavage by briefly centrifuging them. These eggs (called **coenocytic eggs)** continue to undergo nuclear division even though cytoplasmic division does not occur. Activation of transcription in these coenocytic eggs occurs at the same time as in control embryos. The plasmids injected into these eggs become transcriptionally inactive shortly after injection and remain inactive until the midblastula transition-the time when endogenous gene transcription is also accelerated.

Newport and Kirschner reasoned that if the amount of DNA is the critical factor that determines when transcription is activated, transcription of the plasmids might be activated prematurely if the amount of DNA were increased experimentally. Thus, they injected large amounts of an unrelated plasmid into the eggs to increase the total amount of DNA. Since these coenocytic eggs have not cleaved, this increases the ratio of DNA to total cytoplasm substantially. Transcription of the yeast tRNA genes was reactivated after this injection-well before the midblastula transition. This result suggests that activation of transcription at the midblastula transition is, indeed, dependent upon the accumulation of a critical amount of DNA.

The large size of the Xenopus egg facilitates microinjection, which makes it possible to conduct the kinds of experiments we have just discussed. The ability to monitor transcription of exogenous genes in the living embryo provides a powerful tool for testing hypotheses concerning regulation of transcription. Additional experiments of this type may eventually lead to the identification of the various components that exert regulatory control over gene transcription during early development.

GENE EXPRESSION DURING INITIAL DEVELOPMENT OF MAMMALS

The eggs of most animals are macroscopic and contain relatively large amounts of yolk and cytoplasm, which sustain the embryos until they have reached the stage where they can feed for themselves. The mammalian egg, on the other hand, is small and relatively poor in yolk and cytoplasm. During development the embryo receives nutrients directly from the mother, thereby reducing the embryo's reliance upon its own stored nutrients. These unique aspects of mammalian development require that it be discussed separately.

In this discussion we are utilizing data gathered primarily from studies on rabbit and mouse embryos. This approach produces an oversimplification of mammalian development, since mammals are a very diverse phylogenetic group. However, these are the only mammals for which significant information is available.

The choice of these two organisms by the majority of investigators has not been accidental. Both are common laboratory animals that are easy to maintain, and techniques are available for obtaining large numbers of embryos by hormonal superovulation and for culturing embryos *in vitro* for experimentation. The mouse offers the added advantage that a great deal of information on its genetic characteristics has been obtained over the years. The rabbit embryo presents an advantage over the mouse embryo for many kinds of biochemical and molecular studies: Its diameter is approximately twice that of the mouse embryo, and in addition, it undergoes a tremendous increase in size during the blastocyst stage. At the beginning of this stage the rabbit embryo consists of 100 to 130 cells, but it expands to 80,000 cells before implantation. This contrasts with the mouse blastocyst, which consists of only about 60 cells at the time of implantation. The increase in cell number in rabbit development is true growth, in contrast to the increase in cell number during cleavage of nonmammalian embryos, in which the increases in cell number are at the expense of cell size. Blastocyst expansion provides investigators with more material for conducting biochemical and molecular studies. These and other differences between the two embryos indicate that they utilize quite different developmental programs. Thus, we cannot extract a concept of typical mammalian development from either species. The final picture of mammalian development is likely to be a composite of several different modes of differentiation, and it remains to be seen which differences are significant and which are trivial.

Gene expression during early mammalian development is of more immediate consequence than it is for the nonmammalian embryos that were discussed earlier in this chapter. This reliance of the early mammalian embryo on expression of embryonic genes is evident when fertilized eggs are treated with inhibitors of RNA synthesis. Although actinomycin D has been used in a number of these studies, it has several side effects on embryos that are unrelated to its inhibitory effects on transcription. Accordingly, a-amanitin is considered to be a more specific inhibitor of transcription. Treatment of mouse and rabbit embryos with this inhibitor causes development to be halted during cleavage. Cleavage arrest is not immediate, however, since a limited amount of cleavage occurs in the presence of α-amanitin. This delay is particularly pronounced in the rabbit embryo; zygotes will develop to the eight-cell stage before development is arrested. The concentration of a-amanitin used in these experiments (10^{-4}M) is sufficient to virtually abolish RNA synthesis immediately, but the effects on protein synthesis are delayed.

These results lead to two conclusions. First, the delayed effect of the inhibitor on protein synthesis and cleavage indicates that some oogenic messenger is translated during the initial cleavages. Second, embryonic gene expression is essential for the continuation of cleavage. Thus, cleavage is maintained by stored oogenic transcripts as well as by new transcripts synthesized from the genomes of embryonic cells. This points out a major difference between mammals and nonmammalian species: Cleavage in the latter group does not depend upon new transcripts but can occur in the absence of *de novo* transcription. However, it is necessary to point out that although the oogenic transcripts in the rabbit egg are of little consequence in terms of the *extent* of development they will support, they are certainly significant in the *length of time* during which they are capable of exerting an influence on the embryo. Oogenic RNA will support development of the mouse zygote to the twocell stage and the rabbit zygote to the eight-cell stage, which is a period of 24 hours in both cases. This is a relatively long time in developmental terms.

The results of inhibitor studies are supported by the results of experiments with manually enucleated mouse eggs. In these experiments unfertilized and fertilized eggs were enucleated and cultured in a medium containing ^{35}S-methionine. The pattern of protein synthesis in the enucleated eggs was compared to the pattern in control embryos by examining the array of radioactive polypeptides on two-dimensional electrophoresis gels. The enucleated fertilized eggs synthesized a number of polypeptides that are characteristically synthesized at the early two-cell stage of embryonic development. However, unfertilized enucleated eggs failed to synthesize these characteristic polypeptides. These results suggest the presence of oogenic messenger in mouse eggs. The difference in the protein synthetic patterns of unfertilized and fertilized enucleated- eggs indicates that utilization of certain of these transcripts can occur only after fertilization.

The continuation of cleavage and specific protein synthesis in the absence of *de novo* RNA synthesis implies that the unfertilized mammalian egg contains a store of ribosomes, tRNA, and messengers. The presence of oogenic messenger RNA in unfertilized rabbit and mouse eggs is also suggested by the detection of polyadenylated RNA in them. Although not absolute proof of their function as messengers, the presence of poly (A) tracts on the RNA molecules is compatible with this interpretation. A large proportion of this polyadenylated RNA is found associated with ribosomes in unfertilized eggs, indicating that it is functioning as messenger RNA. The majority of the ribosomes in unfertilized mouse eggs are, however, not active in protein synthesis and are presumably stored to support postfertilization protein synthesis.

The presence of oogenic messengers in unfertilized eggs is confirmed by in *vitro* translation of mRNA extracted from unfertilized mouse eggs and by RNA blot analysis, which demonstrates the presence of *specific* messenger sequences. Braude et al. (1999) translated messenger RNA isolated from unfertilized mouse eggs. A large number of polypeptides are encoded by these messengers. The polypeptides that are synthesized by intact unfertilized eggs and early two-cell-stage embryos, respectively. Note the polypeptides indicated by the three arrows. These three polypeptides are synthesized in embryos but not in eggs. Braude et al. (1999) have shown that they are also synthesized in o-amanitin-treated embryos, which implies that they are produced from oogenic messengers. The messengers for these polypeptides are, indeed, present in the oogenic messenger RNA. Their translation only after fertilization indicates that their utilization is under posttranscriptional control during development.

RNA blot analysis is a powerful tool for the demonstration of specific RNA sequences. Giebelhaus et al. (1983) have used this technique to demonstrate the presence of both actin and histone H3 mRNA in unfertilized mouse eggs. As additional probes become available, RNA blot analyses with them should eventually result in the emergence of a catalog of sequences in the mammalian oogenic messenger population.

Transition Of Control Over Development From Oogenic To Embryonic Transcripts

The necessity of nuclear intervention during early mammalian development, which is indicated by the inhibitor studies discussed earlier, is supported by a growing body of genetic evidence obtained from the study of mouse mutants. In these experiments, mutants that affect the primary structure of proteins have been used to estimate the time of onset of embryonic gene expression during mouse development. Males and females that differ genetically are mated, and embryonic gene expression is indicated by detection of paternal polypeptides, usually by electrophoresis. The earliest detection of a paternal polypeptide in such an experiment with mouse mutants is the synthesis of β-microglobulin at the two-cell stage. Clearly, embryonic gene expression occurs very early in the mammalian developmental program.

Analyses of RNA synthesis during mammalian development indicate that transcription occurs from the very beginning of development. Mintz (1984) was the first to demonstrate RNA synthesis in early mammalian cleavage stages. Mintz's autoradiograph of a twocell mouse embryo, which has been labeled with ^{3}H-uridine. Incorporation of the label into RNA is clearly evident. Characterization of the classes of RNA synthesized during mouse development indicates that DNA-like heterogeneous RNA is synthesized as early as the two-cell stage. This RNA is likely messenger RNA, an interpretation that is reinforced by the detection of newly synthesized polyadenylated RNA in two-celled mouse embryos. Ribosomal and transfer RNA synthesis are apparently delayed somewhat. There is an indication of rRNA synthesis at the two-cell stage, and its synthesis has been unequivocally demonstrated at the four-cell stage. The four-cell stage also marks the first appearance of newly synthesized 4S RNA.

The cleavage-stage mouse embryo possesses transcripts of dual origin. However, a sustained impact of oogenic transcripts on development is tempered by degradation of these transcripts. The loss of oogenic transcripts is particularly striking during the two-cell stage when the amount of polyadenylated RNA falls dramatically. This is followed by a buildup of this category of RNA between the late two-cell and early blastocyst stages. The latter, of course, results from embryonic gene transcription. The shift in the source of transcripts apparently results in modifications to the messenger population that are sufficient to produce changes in the pattern of protein synthesis. This conclusion is based upon the observation that the appearance of a distinct set of proteins at the late two-cell stage is prevented by treatment of the embryos with a-amanitin. Before this stage, of course, a-amanitin does not have a significant effect on the protein synthetic pattern.

The behavior of particular sequences during the loss of oogenic messengers and replacement by embryonic messengers can now be analyzed by messenger hybridization to cloned probes. As we mentioned in the previous section, the presence of histone H3 and actin sequences in unfertilized eggs has been demonstrated by RNA blot hybridization. The changes in the amounts of these sequences during development have also been determined. In both cases the sequences are present in substantial amounts in the egg, decrease at the two-cell stage, and show a progressive increase from the eight-cell to blastocyst stages. The degradation of oogenic messenger before accumulation of embryonic messenger is reminiscent of the situation for actin messengers in the sea urchin *Strongylocentrotus drob*achiensis.

The data presented in this and previous sections of this chapter demonstrate that RNA synthesis occurs during cleavage of both mammalian and nonmammalian embryos. The major difference between these groups appears to be the *extent of involvement* of the newly synthesized RNA in development. In nonmammalian species, development will continue in the absence of transcription to a much more advanced stage than it will in the mammalian embryo. Thus, the embryonic genome intervenes in development of mammalian embryos during early cleavage, whereas cleavage of nonmammalian embryos is directed primarily by the oocyte genome.

X Chromosome Gene Expression

The role of the genome in developmental regulation requires that the input of genomic information into the protein synthetic machinery be precisely regulated. A regulatory mechanism in female mammals that modulates the utilization of X chromosome information; this mechanism is dosage compensation. Female mammals have two X chromosomes, whereas males have but one. Dosage compensation ensures that only one X chromosome is active in cells of the female. According to the Lyon hypothesis, one of the X chromosomes in the cells of a female becomes inactivated during development, and hence unavailable for transcription. The inactivated X becomes heterochromatic and is consequently highly condensed during interphase (the Barr body) and late replicating. The inactivation is heritable; every mitotic derivative has the same X inactivated. The initial inactivation is generally random; that is, either the paternal or maternal X may be inactivated in any cell. The Lyon hypothesis therefore predicts that a discrete regulatory event in embryonic cells inactivates one X and leaves the other transcriptionally active. Accordingly, both the maternal and the paternal X chromosomes should be euchromatic before this event. Since the embryonic genome is utilized early in mammalian development, it is also possible that both are transcriptionally active before inactivation.

Cytological evidence indicates that both X chromosomes are euchromatic during early cleavage, since neither X is condensed nor late replicating. A late-replicating X chromosome is first detected in the late blastocyst stage of mouse development. Furthermore, one of the X chromosomes shows differential staining at approximately the same time-an indication of heterochromatization. Therefore, it is possible that both X chromosomes are transcriptionally active during cleavage, although we should consider the possibility that functional inactivation precedes cytological inactivation. If both X chromosomes are active in the female embryo, she should produce twice as many proteins specified by X chromosome genes as her brothers.

In an elegant experiment Epstein et al. (1998 have examined the activity of hypoxanthine-guanine phosphoribosyl transferase (HGPRT), an X-linked enzyme, during the early blastocyst stage. To determine the sex of the blastocysts, Epstein and his associates used half embryos that had developed from intact embryos separated at the two-cell stage. The twin half embryos were cultured to the blastocyst stage, and one half was used for HGPRT assay, whereas the other half was used for determination of sex by chromosome analyses. A comparison of HGPRT activity between male and female early blastocysts. The ratio of female to male HGPRT activity from the combined data is 2.2:1.0. These results are consistent with the interpretation that both X chromosomes are active up to the early blastocyst stage. In additional experiments the investigators determined the enzyme activities of late blastocysts and found no difference between male and female embryos. This result suggests that X inactivation had occurred between the early and late blastocyst stages. A similar result has also been obtained by Kratzer and Gartler (1998).

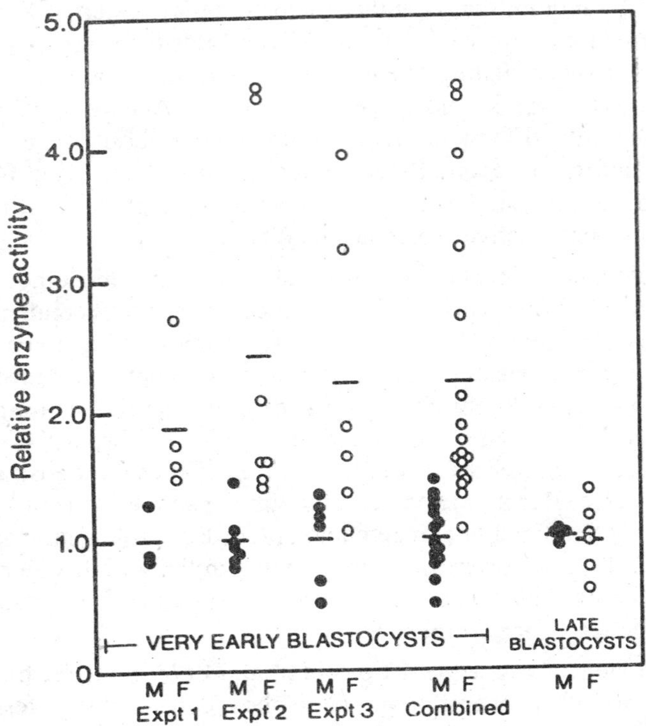

Figure 7.8 : Relative activities of hypoxanthine-guanine phosphoribosyl transferase (HGPRT) in male and female blastocysts. After determination of sex, the individual half-blastocysts were pooled into groups of five of the same sex and were assayed for enzyme activity. To facilitate comparisons among experiments, the results of each activity were normalized to a mean male enzyme activity of 1.0. M: male; F: female.

Another means of determining whether both X chromosomes are active during early development is by cell transfer experiments. These experiments were conducted with embryos from a mouse strain heterozygous for X-linked pigmentation mutants. Single cells derived from blastocysts of this strain were injected into host blastocysts, which developed into chimeric mice with patches displaying the pigment of both X-linked genes of the donor cell. Since both kinds of pigment patches were derived from *a single donor cell,* it is obvious that X inactivation occurred in descendants of the donor cell *after* cell transfer.

Although we have described X chromosome inactivation in the embryo as being random, there is evidence that in certain extraembryonic tissues of the rodent embryo the maternal X chromosome is preferentially expressed. This results from nonra-ndom inactivation of the paternal X chromosome in the cells of these extraembryonic tissues. Preferential behavior of the X chromosomes in extraembryonic cells suggests that the maternal and paternal X chromosomes are somehow imprinted to indicate their origins. However, the imprinting of the X chromosomes is inconsequential for X inactivation in embryonic cells.

MACROMOLECULAR SYNTHESIS DURING SEED DEVELOPMENT AND GERMINATION

The angiosperm zygote begins its development within the ovule. When the embryo is fully formed, it is enclosed by a seed coat, which is derived from the outer integument of the embryo sac. The embryo, nutrient storage tissues, and the seed coat constitute the seed.

At the completion of embryonic development in many plants, there is a water loss from seed tissues, which leads to desiccation of the embryo cells. The embryo then enters a **quiescent period,** during which development and metabolic activity cease until the seed germinates. For some seeds, the only requirement for germination processes to commence is imbibition of water. This is accompanied by a resumption of metabolic activity. In other seeds, imbibition of water is also necessary for metabolism to commence, but it is insufficient to trigger germination; additional environmental factors or internal changes are required. In the latter situation the ungerminated seeds are said to be **dormant.** Germination is completed when elongation of the embryonic axis leads to the rupture of the seed coat and protrusion of the elongating radicle from the seed; this is the commencement of seedling growth. In the discussion that follows our primary concern will be with the macromolecular events occurring during seed development and germination so as to foster an appreciation for the utilization of genomic information during these fascinating phases of plant development.

Gene Expression During Seed Development

Macromolecular synthesis during the interval between fertilization and the onset of quiescence has two roles: in the construction of the seed itself during development and in facilitating the rapid reinitiation of activity upon subsequent germination. As we shall see, the latter process is partially dependent upon transcripts that are produced during embryogenesis and stored in the seed for utilization after imbibition.

As with animal development, the study of gene expression during plant development is facilitated by monitoring the expression of genes encoding proteins that are developmentally regulated; that is, genes that are expressed either during a restricted phase of the life cycle or in particular tissues. Excellent candidates for such an analysis are genes encoding the seed proteins-proteins made in large amounts only during embryonic development. These proteins accumulate during seed **maturation.** This is the phase of seed development after the embryonic axis has been formed when cell division has ceased and rapid cell enlargement occurs. The cotyledons show extensive growth during this phase. Among the seed proteins that are synthesized during maturation are the **seed storage proteins,** which accumulate primarily in the cotyledons and are the predominant proteins in the mature seed. In the soybean, for example, storage proteins constitute approximately 70% of total seed protein and account for 30 to 40% of seed weight. Storage proteins are synthesized on the rough endoplasmic reticulum and deposited in membrane-enclosed protein bodies.

After seed maturation there is a drastic reduction in metabolic activity as the embryo and its protective seed coat desiccate. After desiccation the seed is shed from the plant and remains quiescent until germination; during subsequent seedling growth the protein bodies containing the storage proteins are catabolized and used as a food source.

An advantage for studying gene expression during maturation is that this process can be controlled during *in vitro* culture. Normally, immature embryos removed from developing seeds do not mature when cultured; they germinate precociously. However, addition of the growth regulator **abscisic acid (ABA)** will prevent this precocious germination and allow maturation to proceed. ABA production during embryogenesis is thought to be the natural trigger of seed maturation and quiescence. In some plants ABA production is stimulated by water stress. Therefore, as the seed begins to dry, ABA is produced, the seed matures and becomes quiescent.

As a consequence of their exposure to ABA, cultured immature embryos undergo a number of modifications in macromolecular synthesis that are characteristic of the maturation phase of development. For example, synthesis of seed proteins (including storage proteins) can be induced by addition of ABA to the culture medium. The discovery of a regulator of the maturation phase of the life cycle with effects on the synthesis of embryo-specific proteins is an exciting one because of the potential for the analysis of the control over gene expression. Additional research should reveal whether ABA regulates gene expression directly and, if so, at what level(s) in information utilization.

The stage specificity of synthesis and the abundance of seed storage proteins make the analysis of expression of the genes encoding them an excellent system for developmental analysis. As we have learned, abundant proteins are products of translation of prevalent messengers, which present many technical advantages. Their patterns of synthesis and accumulation are relatively easy to monitor. Recently, cDNAs have been prepared to the soybean storage protein messengers; these cDNAs have been cloned and have become valuable probes of the transcripts. Furthermore, the cDNAs have been subjected to nucleotide sequence analysis and used as probes for complementary sequences in the genome.

Recent RNA blot analyses with cloned soybean storage protein cDNAs demonstrate that polyadenylated RNA that hybridizes to these cDNAs is missing until the early cotyledon growth phase, increases dramatically in amount during cotyledon growth, and is reduced below the level of detection in mature, desiccated seeds. Thus, accumulation of these messengers is developmentally regulated. They are most abundant when the seeds are active in storage protein synthesis, suggesting that the primary control over storage protein synthesis is at the transcriptional level. Transcr-iptional-level regulation of the structural genes for abundant soybean seed proteins (including the storage proteins) has been confirmed by Goldberg et al. (1981b). Hybridization analysis with cloned cDNAs to the messengers for these proteins shows that the messengers accumulate during embryogenesis and decay before seed hydration. Furthermore, transcripts that hybridize to the DNA probes are missing from both the cytoplasmic and nuclear RNA of leaf cells. Thus, these genes are transcriptionally silent except during a short period in embryogenesis when they produce massive amounts of messenger primarily in cotyledon cells. They are analogous to genes in animals that encode abundant, developm-entally regulated cell-specific proteins, such as globin and ovalb-umin, whose genes are also regulated at the transcriptional level. Since the synthesis of the abundant seed proteins can be induced with ABA in many kinds of seeds, it will be interesting to learn whether transcription of their structural genes is regulated by this growth regulator.

In addition to the highly abundant seed protein messengers, a variety of other messenger RNAs are synthesized and accumulate during embryogenesis. DNA-RNA hybridization experiments indicate that throughout their development, soybean embryos contain a nearly constant population of 14,000 to 18,000 diverse messenger RNAs. In the final stage of seed development, polysomes decay as messengers and ribosomes dissociate. Most of these mRNA sequences, however, persist in the dry seed. As we shall discuss in the next section, these messengers stored in the quiescent seed are utilized for protein synthesis during germination. In contrast to the highly abundant seed protein messengers, these diverse sequences are present not only throughout embryogenesis and in the quiescent seed but are also detectable in leaf- polysomes. Thus, there appear to be two distinct categories of structural genes that are transcribed during embryogenesis: a small number encoding highly abundant seed protein messengers that are transcribed during a limited phase of embryonic development and a vast number of genes that are presumably encoding generalized proteins needed in small amounts throughout the life cycle and not subject to transcriptional modulation.

The Control of Protein Synthesis During Germination

Immediately upon imbibition ribosomes associate with messenger RNA and translation begins. The necessity for protein synthesis during germination is illustrated by treatment of imbibing wheat seeds with cycloheximide; protein synthesis is inhibited and germination is prevented. Experiments with wheat seeds indicate that the initiation of protein synthesis does not require *de novo* RNA synthesis. When more than 90% of RNA synthesis is prevented with a-amanitin, polysome formation is not significantly reduced during the first 40 minutes of imbibition. This result is interpreted to mean that early protein synthesis is directed by long-lived messengers that are synthesized during embryogenesis, stored in the seed, and recruited during imbibition. Translation of these messengers must utilize stored ribosomes and tRNA.

After a short period of reliance on preformed messengers, responsibility for protein synthesis shifts to newly synthesized messengers. This shift of responsibility results in changes in the pattern of protein synthesis. The changes in the pattern of protein synthesis are eliminated when the imbibing seeds are exposed to α-amanitin. This result suggests that the changes in the pattern of protein synthesis are dependent upon de novo messenger RNA synthesis. This interpretation is strengthened by the results of *in vitro* protein synthesis: Messengers isolated from late postimbibition embryos direct the synthesis of proteins resembling those formed during early imbibition, whereas those isolated from late postimbibition embryos direct the synthesis of proteins resembling those formed during late imbibition. Thus, as in postfertilization development of animals, a supply of preformed messengers directs protein synthesis until those messengers are degraded and are replaced by newly synthesized messengers.

8

Nucleo-Cytoplasmic Interactions

We have repeatedly seen that the nucleus and the cytoplasm continuously interact during normal egg development. These interactions can be experimentally modified in two different ways. The most radical experiment is to divide the egg into two parts (merogony) and to compare the morphogenetic and biochemical potentialities of the *nucleate* and *anucleate* halves. Another approach is to modify the *nucleus.* One of the simplest ways to attain this goal is to introduce a foreign nucleus into the egg by hybridization or nuclear transplantation.

Biochemical work on anucleate fragments of eggs and on hybrids will be the main topic of this chapter. We shall, however, start with another object, the unicellular giant alga *Acetabularia,* since this organism presents a number of advantages for those who are interested in the role of the nucleus in morphogenesis and in the biochemical interactions between the nucleus and the cytoplasm.

BIOLOGY AND BIOCHEMISTRY OF THE ALGA ACETABULARIA

Biological Cycle. Regeneration

The giant green alga, *Acetabularia mediterranea,* which is 3 to 5 cm long, has a single nucleus during the major part of its life cycle. This nucleus, which is large and contains giant RNA-rich nucleoli, but very little chromatin, is localized at the basal part of the alga (which is called the *rhizoid)*. At the apical end of its stalk, the still uninucleate alga forms an umbrella (or *cap)* which will later serve for its sexual reproduction (formation of cysts, which will produce motile gametes). The morphology of the cap is species-specific and allows taxonomists to classify the different species of *Acetabularia.* The most important of these species, for biochemists, are *A. mediterranea* and *A. crenulata,* which can easily be cultivated in large amounts in the laboratory.

Morphogenesis is affected, in *Acelabularia,* by the amount of *light* it receives. If the light supply is abundant, caps form quickly on short stalks. Reduced light supply results in the formation of very long stalks and delayed production of small caps. Absence of light completely stops growth and induces a number of degenerative changes in the nucleus (decrease in size, reduction of the dimensions and RNA content of the nucleolus); all these changes are reversible if light is again supplied to the algae. These simple experiments clearly show that the structure and chemical composition of the nucleus depend on the energy supply in the cytoplasm derived from photosynthesis.

Acetabularia has become a fascinating subject for biologists and biochemists since the fundamental work of Hammerling around 1930. He discovered that if the alga is cut into two pieces, both the nucleate and anucleate fragments can *regenerate* and form a typical cap. Anucleate fragments can survive for several weeks and form caps, provided that the apical part of the stalk is present. The algae thus contain *morphogenetic substances,* which are distributed along an *apico-basal gradient;* basal fragments, although they are close to the nucleus, hardly regenerate and never form caps. Nevertheless, the morphogenetic substances are *gene products* and originate from the *nucleus.* This has also been demonstrated by Hammerling, who succeeded in grafting nucleate fragments of A. *mediterranea* on anucleate stalks of A. *crenulata* and vice versa. The result is, in general, the formation of a "hybrid" cap, which soon degenerates and is replaced by the typical cap of the nucleate fragment. There is a kind of "struggle" between

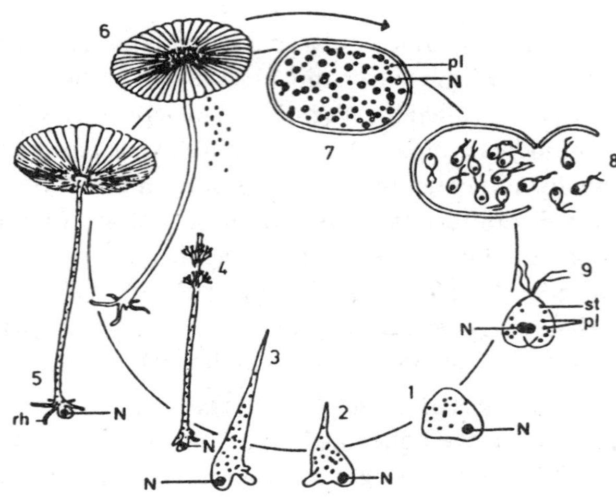

Figure 8.1 : Acetabularia mediterranea. Upper left: Life cycle of the alga. 1 zygote; 2 young growing cell; 3 slowly growing alga; 4 alga in its rapid growing phase; 5 alga with a young cap; 6 alga with a mature cap; the vegetative nucleus N has broken down and cysts disperse out of the cap. 7 Enlarged resting cyst; 8 germinating cyst giving rise to the gametes; 9 conjugation between two gametes. N nuclei; pl chloroplasts; st stigma; rh rhizoids.

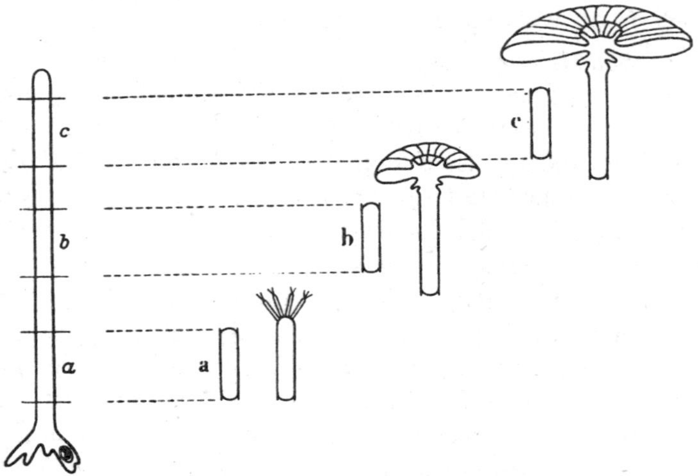

Figure 8.2 : Distribution of morphogenetic substances according to the experiments of J. Hammerling.

the pre-existing morphogenetic substances present in the anucleate half of one species and the newly formed substances produced by the nucleus present in the nucleate half. The battle is ultimately won by the nucleus. The reasons why the gradient in the distribution of the morphogenetic substances decreases from the apical toward the basal end of the stalk and not, as one would expect, from their nuclear origin, in the opposite direction, are not clear. Experiments by Russian workers have shown that it is possible to displace, by centrifugation, almost the whole content of the alga toward the apical or the basal end of an anucleate fragment. Nevertheless, the polarity remains unchanged and the cap forms at the initial apical end. Thus, it is probable that polarity is linked to the properties of the cell membrane and cell wall, which are different at the apical end, where growth occurs. There is an accumulation of RNA, proteins, glycoproteins and polysaccharides at this apical end. It is possible, but this remains a hypothesis, that the membrane contains specific receptors for the morphogenetic substances; such receptors would be accumulated at the apical end of the stalk.

Morphology of the Cytoplasm and the Nucleus

The algae are protected by a thick extracellular wall made of polysaccharides; it is in close contact with the plasma membrane which limits the cytoplasm. The center of the alga is occupied by a large, turgescent, acidic vacuole. The cytoplasm is reduced to a thin sheet, compressed between the extracellular membrane and the central vacuole. Golgi bodies, which are accumulated at the growing apical end of young algae where they participate in the formation of the polysaccharide cell wall, and mitochondria are abundant in the cytoplasm. But the most obvious cell organelles are the *chloroplasts,* they are ovoid and often contain inclusions of the polysaccharides, which they have synthesized by photosynthesis. If one observes an *Acetabularia* under the light microscope, one can see that the chloroplasts are moving continuously on cytoplasmic strands; this movement is called *cyclosis.* Cytoplasmic contractile proteins (actin, myosin) are involved in cyclosis, since chloroplast movement quickly stops if the algae are treated with the actin-binding drug, cytochalasin.

The large (300.tm in diameter) *"vegetative" nucleus* of the alga contains, according to the *Acetabularia* species, a single large ribbon-shaped nucleolus or several smaller nucleoli. The nucleoli are very rich in RNA and are the site, as we shall see, of intensive rRNA synthesis. It has been impossible for a longtime to demonstrate cytochemically the presence of DNA in the huge vegetative nucleus. On the other hand, DNA is easily detectable in the much smaller nuclei of the gametes and zygotes. Recent studies have shown that the vegetative nucleus contains *lamp brush chromosomes* similar to those of amphibian oocytes: 20-µm-long, DNA-containing loops extend in the abundant nuclear sap from condensed chromomeres. The nucleus, despite its large size, is diploid and contains a little more DNA than the theoretical diploid value, which is due, as we shall see, to amplification of extrachromosomal (nucleolar) genes. When the cap reaches its full size, the nucleus shrinks, the nucleoli regress and a huge intranuclear spindle is built up. Meiosis occurs at this time. The numerous small daughter nuclei, which move toward the cap and invade it, are haploid. In the cap, groups of such "secondary" nuclei, which have been repeatedly dividing, become surrounded by a polysaccharide membrane: the cap is now subdivided in cysts; each of them contain many haploid nuclei. Finally, the cysts are released in seawater and give rise to a swarm of flagellated haploid gametes.

As one can see, there is a remarkable similarity between *Acetabularia* and eggs from animals so far as the behavior of the nucleus is concerned: the vegetative nucleus is, like the oocyte germinal vesicle, in meiotic prophase. In both cases, the nucleoli are RNA-rich and well-developed; the chromosomes are in an extended lampbrush state and the nuclear sap is abundant. However, the life cycles diverge after meiosis: in *Acetabularia,* the repeated cell divisions of the secondary nuclei precede the formation of gametes and zygotes; in eggs, cleavage follows fertilization.

Biochemical Studies

Morphogenetic Substances and mRNAs: Experiments with Inhibitors

The main problem is, of course, the *chemical nature of the morphogenetic substances.* There are many good reasons for believing that since they are synthesized in the nucleus and carry the information necessary for the production of the many proteins which are certainly needed for cap formation, the morphogenetic substances are *a family of stable mRNA molecules. No* experiments contradicting this hypothesis have so far been published and there is, as we shall see, ample circumstantial evidence in its favor. However, direct proof, such as isolation of the mRNAs produced by the nucleus and demonstration that they have biological activity, is still missing.

In favor of the mRNA hypothesis are a number of facts. *UV-irradiation,* at wavelengths which affect the integrity of nucleic acids, inhibits regeneration in anucleate halves, in particular when the apex of the stalk, where morphogenetic substances are accumulated, is irradiated. Irradiation of nucleate halves has only a transitory effect. Regeneration is stopped and then resumes when the nucleus has been damaged in a reversible way only. Similar results are obtained when living fragments of the algae are treated with the enzyme *ribonuclease. Loss* of regeneration is irreversible after treatment of anucleate fragments, reversible after action of ribonuclease on the nucleate ones. Ribonuclease apparently destroys the pre-existing mRNAs; new mRNAs cannot be produced unless the nucleus is present.

The mRNA hypothesis predicts that, on the contrary, actinomycin should not .inhibit the regeneration of the anucleate halves (since the drug does not bind the pre-existing RNAs), but would arrest that of nucleate halves (because actinomycin will inhibit the synthesis of new mRNA molecules by the nucleus). These expectations are fulfilled by experiments.

In actinomycin-treated anucleate fragments, an unexpected secondary effect is, however, observed after a few days. While the initiation of cap formation was as good as in the controls, the growth of the anucleate caps is definitely depressed. The probable explanation of this secondary effect is the following: the *chloroplasts,* which exist in very large numbers in the algae, contain their own DNA. Actinomycin binds to chloroplastic DNA and this binding results in alterations of chloroplast activity which lead to a slowing down of the outgrowth of the cap.

Autoradiography brings further evidence for the view that the morphogenetic substances are stable mRNAs. If a normal alga is treated for a short time with labelled uridine, this precursor is very quickly incorporated into nuclear RNA. If the algae are cultivated, after this uridine "pulse", in normal seawater, one can follow the migration of the RNAs which have been synthesized in the nucleus. They first move into the cytoplasm, then accumulate at the apex of the stalk (where, as we know, the morphogenetic substances are also concentrated). We shall soon see that more recent biochemical evidence also support the stable mRNA hypothesis.

Energy Production

Oxygen consumption and *photosynthesis* remain normal in anucleate fragments for a few days. Afterward, the rate of photosynthesis somewhat decreases, as well as the ATP content, in the anucleate halves.

Of particular interest is the existence of a *circadian rhythm* in photosynthetic activity. If the algae are submitted to alternate periods of light and darkness, each period lasting about 12 h, they "remember" this photosynthetic rhythm for a certain time when they are cultivated in continuous light. Anucleate fragments retain their circadian rhythm of photosynthesis for many weeks. If however, nucleate fragments with a given rhythm are grafted on anucleate halves which have a different periodicity in their rhythm, the nucleus imposes its rhythm on the chloroplasts of the anucleate halves. It is even possible to combine, in such grafting experiments, nucleate halves which have lost their rhythmicity with anucleate fragments which possess a normal rhythm or vice versa. Only when the nucleate half has retained its rhythm will the graft be the site of a circadian rhythm.

This, as well as other experiments made with inhibitors of RNA and protein synthesis, has shown that the circadian rhythm of photosynthesis is not completely autonomous in anucleate halves. A nuclear control, probably exerted by mRNAs, superimposes itself on cytoplasmic rhythmicity and controls it.

Similar results and conclusions have been obtained for several other circadian rhythms found in *Acetabularia*. They deal with the shape of the chloroplasts, the RNA, polysaccharide and ATP content, etc.

Protein Synthesis

The protein content of both nucleate and anucleate halves approximately doubles during the 2 weeks which follow the sectioning of the algae. Since many of the newly synthesized proteins are specific *enzymes* and since we know that enzyme synthesis is controlled by the genes, it follows that the information which emerges from the genes must be stored in the cytoplasm of the anucleate fragments. The only explanation that molecular biology can offer for such a paradox is the intervention of stable mRNA molecules, which have been synthesized in the nucleus and stored in the cytoplasm.

Certain enzymes, for instance, the phosphatases and the enzymes involved in the production of the cell walls, markedly increase in activity when the caps form. This increase occurs in anucleate as well as in nucleate fragments. Some of these enzymes are distributed along the apico-basal gradient as morphogenetic substances, a fact which suggests that the corresponding mRNAs might have the same spatial distribution.

Regeneration, in anucleate fragments, is quickly and irreversibly blocked by *puromycin* and *cycloheximide,* the two classical inhibitors of cytoplasmic protein synthesis. On the other hand, *chloramphenicol* and *tetracycline,* which inhibit chloroplastic and mitochondrial protein synthesis, have very little effect on regeneration. Since the effects of puromycin and cycloheximide on regeneration are reversible in the case of nucleate fragments, one can safely conclude that proteins synthesized on cytoplasmic polysomes, under the direction of mRNAs of nuclear origin, play a more important role in morphogeneis than the newly synthesized chloroplastic proteins.

The chlorophasts contain their own enzymes and it would seem likely, at first sight, that this synthesis is controlled by chloroplastic rather than nuclear DNA. Experiments by Schweiger, in which nucleate halves of A. *mediterranea* were grafted into anucleate halves of *A. crenulata* and vice vers, have shown that this is not true for the lactic and malic dehydrogenases. The enzymes present in the chloroplasts of the "hybrid" progressively change and are replaced by others, which are characteristic of the species which provided the nucleus. The same nuclear control has been found for the synthesis of the insoluble proteins which form the chloroplastic membrane and of the chloroplastic ribosomal proteins.

Inhibitors which discriminate between chloroplastic and cytoplasmic protein synthesis have been very useful for the analysis of enzyme synthesis in *Acetabularia.* As we have seen, puromycin and cycloheximide inhibit the functioning of the cytoplasmic 80S ribosomes, while chloramphenicol blocks protein synthesis in the chloroplastic 70S ribosomes and in the still smaller mitochondrial ribosomes. Use of these inhibitors has disclosed an unexpected fact: one would have expected that the activity of the enzymes involved in DNA synthesis increases when the vegetative nucleus breaks down and the daughter nuclei multiply at a fast rate. This is indeed what happens for thymidine kinase and other enzymes important for DNA synthesis; but a surprise came when it was found that these enzymes are encoded by chloroplastic DNA, not by nuclear genes.

Subtle interactions between the nuclear and chloroplastic genomes thus take place during the life cycle of the alga, in particular at the time of cap formation. In order to throw some light on these interactions, a direct approach has been taken recently by H.G. Schweiger and colleagues. They injected mRNAs into the algae and studied their in vivo translation on 80S cytoplasmic ribosomes. The main conclusion was that during the *Acetabularia* life cycle, translation of some proteins is not regulated, i.e. it proceeds for months at a constant rate; but translation of other proteins is turned off at a given stage. In contrast, at the same stage, translation of still another group of proteins is enhanced. Continuation of these experiments should throw more light on the complex pattern of protein synthesis regulation during growth and morphogenesis in *A cetabularia.*

RNA Synthesis

The total RNA content, like the protein content, doubles within 2 weeks when caps are formed in either nucleate or anucleate fragments. This large increase is mainly due to the synthesis of *chloroplastic* RNAs. Only recent work, in the laboratories of H.G. Schweiger and W. Franke, has allowed precise studies on RNA production by the nucleus.

Amplification of the *ribosomal* (nucleolar) genes occurs, as in amphibian oocytes, in the *Acetabularia* vegetative nucleus. Electron microscopy of spread nucleoli and estimations of their DNA content have shown that there are about 4000 copies of the 28 S and 18 S rRNA genes in the vegetative nucleus. As in the oocyte nucleus,. this number of copies does not increase markedly during the growth period of the vegetative nucleus, thus, amplification apparently takes place soon after zygote formation.

The general organization of the ribosomal genes is very similar in the nucleolar organizer of *Acetabularia* and *Xenopus* oocytes. Tandem repeated transcription units are separated by untranscribed spacers of varible lengths. However, these spacers are, on the average, smaller in *Acetabularia* an in *Xenopus* and the size of the nucleolar rRNA precursor, in the alga, is not uch larger than that of the final products (28 S+ 18 S rRNAs). The rate of rRNA synthesis by the vegetative nucleus is very high (4 × 10' nucleotides/ per second per nucleus); it is increased 20 times a few days after removal of part of the stalk. The ribosomal RNAs are quickly transferred from the nucleus to the cytoplasm, where they display a great stability (half-life of 80 days); they move from the rhizoid to the apex of the stalk with a speed of 2-4 mm day. As one can see, the *Acetabularia* nucleus is, like the oocyte nucleus, a machine for the large production of cytoplasmic ribosomes.

Fortunately, the chloroplastic mRNAs have no polyadenylic "tail" and this has given the possibility of selectively "catching" the *messengers* produced by the vegetative nucleus. As expected, these mRNAs of nuclear origin are synthesized only by nucleate fragments of the alga and are heterogeneous: their molecular weights range between 0.5 and 3×10^6. They migrate from the nucleus to the apex of the stalk independently of the rRNAs at a speed of 5 mm day^{-1} (the growth rate of the stalk itself is only 1-3 mm day'), thus independently from the ribosomal RNAs.

The increase in the rate of mRNA synthesis (two- to threefold) after cutting the alga into two halves, is much smaller than for the rRNAs. These data suggest that transcription of chromosomal DNA plays only a limited role in the control of genetic activity and morphogenesis in *Acetabularia.* The major role is played (as in maturing *Xenopus* oocytes and fertilized sea urchin eggs) by the selective translation of mRNAs of nuclear origin stored at the apex of the alga.

Little is known, unfortunately, about the nature of the factors which regulate the translation of the mRNAs accumulated at the apex of the stalk. Inhibitors of *proteolytic enzymes* prevent cap formation in both nucleate and anucleate fragments: it might be that protease activity is required for the "unmasking" of the mRNAs of nuclear origin which are linked to proteins in the form of ribonucleoprotein particles. Another regulatory factor is the *polyamine* content: inhibitors of ornithine decarboxylase (ODC) prevent cap formation in both kinds of fragments. They also prevent the formation of cysts, if they are added to whole algae after the breakdown of the vegetative nucleus. Since the ODC inhibitors decrease the polyamine content of treated cells, it can be concluded that the putrescine, spermidine and spermine levels control both RNA and DNA synthesis in the alga. In polyamine-depleted algae, the decrease in RNA synthesis is followed by inhibition of growth and morphogenesis; the slowing down of DNA synthesis in the secondary nuclei prevents normal cyst formation.

These experiments with ODC and proteolytic enzyme inhibitors have allowed an estimation of the stability of the morphogenetic substances in anucleate fragments of *Acetabularia.* If such fragments are treated with the inhibitors during different lengths of time and then cultured in normal medium, reversibility becomes partial after a 10-day treatment and completely disappears after 2-3 weeks. This is in good agreement with the estimated half-life of the *Acetabularia* messengers (about 10 days) and provides further evidence for the view that Hammerling's morphogenetic substances are a mixture of mRNAs synthesized by the nucleus and stored in the apical cytoplasm.

DNA Synthesis. Relative Autonomy of the Chloroplasts

As we have seen, there is no replication of nuclear DNA during the whole vegetative life of the alga (which lasts several months). Only when the cap has reached its maximal size does the big vegetative nucleus break down and give rise to daughter nuclei which replicate their DNA, divide and are transferred passively to the cap by protoplasmic streaming.

In contrast, *chloroplastic* and *mitochondrial* DNAs are very actively synthesized by both nucleate and anucleate fragments of *A cetabularia.* The total DNA content increases two to three times in both kinds of fragments during the 10 days which follow the sectioning. This came as a big surprise 20 years age, since the very existence of DNA in chloroplasts seemed very doubtful at the time. Now we know that chloroplastic and nuclear DNAs are very different molecules. In *Acetabularia,* chloroplastic DNA is a mixture of large circular and long linear molecules. In addition, small circular molecules, corresponding to replicating chloroplastic DNA, have been observed under the electron microscope (review by Luttke and Bonotto). The use of highly sensitive and very specific stains for DNA allows us to see chloroplastic DNA with a fluorescence microscope, which is possible because the DNA content of the *Acetabularia* chloroplasts is exceptionally high as compared to that of other unicellular algae. Fluorescence microscopy shows that the DNA content of all *Acetabularia* chloroplasts is not the same, i.e. the small, young chloroplasts, which are accumulated at the tip of growing algae, contain much more DNA than the large chloroplasts which surround the nucleus. In fact, many of these basal chloroplasts, which are heavily loaded with carbohydrate reserves, have probably no DNA at all. There is thus a progressive loss of chloroplastic DNA along the apicobasal gradient.

Chloropastic DNA can, in *Acetabularia* as elsewhere, be transcribed and translated. If they are given light, isolated chloroplasts synthesize all kinds of RNAs and a number of proteins (in particular, proteins involved in the formation of chloroplastic membranes and lamellae). However, as we have seen, the synthesis of enzymes such as lactic and malic dehydrogenases and that of the chloroplastic ribosomal proteins requires the co-operation of the nucleus: the *autonomy* of the chloroplasts toward the nucleus is thus imperfect. Counts of the chloroplasts, in nucleate and anucleate fragments of the alga, have shown that they increase in number even in the absence of the nucleus. However, chloroplast multiplication is more active in the nucleate than in the anucleate halves, where many chloroplasts are large and have a dumbbell shape. It is thus probable that cytoplasmic factors, which are synthesized under nuclear control, are required for fission of the chloroplasts and that they are missing in the anucleate halves. Since even an apparently specific chloroplastic function such as the circadian rhythm of photosynthesis is regulated on cytoplasmic (80S) ribosomes, we must again conclude that the chloroplasts are not fully autonomous toward the nucleus. In whole algae, the chloroplastic and nuclear genomes co-operate in order to specify the protein composition of the chloroplasts and to allow their multiplication.

Complexity of Nucleocytoplasmic Interactions

It is clear that the nucleus provides, in *Acetabularia,* the various mRNAs which are required for morphogenesis (cap formation), i.e. the nucleus is the source of the morphogenetic substances and therefore exerts a very important *positive control* on the cytoplasm. It should be emphasized that although anucleate fragments can utilize their stored mRNAs of nuclear origin and form full-sized caps, these caps always remain sterile: neither cysts, nor gametes form in anucleate caps which are therefore useless for reproduction purposes.

But it would be a mistake to believe that the vegetative nucleus of *Acetabularia* exerts only positive effects on the cytoplasm. Already 50 years ago, Hammerling demonstrated that anucleate fragments of *Acetabularia* form caps faster than control whole algae. Similarly, we found that protein synthesis is accelerated when the nucleus has been removed. There is no doubt that the nucleus exerts *negative,* as well as positive controls over the cytoplasm.

In addition, the *cytoplasm* controls nuclear activity. If the photosynthetic function of the chloroplasts is suppressed by culture in darkness, the nucleus shrinks. The nucleolus undergoes vacuolization and stops synthesizing RNA. Thus, energy production in the cytoplasm is absolutely required for the maintenance of nuclear morphology and synthetic activity. If, as shown long ago by Hammerling, one cuts repeatedly an alga into halves when it begins to form a cap, the breakdown of the large vegetative nucleus is inhibited, thus there will be no formation of daughter nuclei, no nuclear DNA synthesis. Repeated amputation of the cap thus leads to immortality. As in amphibian oocytes undergoing maturation, the choice between DNA transcription and replication is decided by cytoplasmic factors. This conclusion is reinforced by more recent experiments in which an old vegetative nucleus was introduced into the cytoplasm of a young alga; this nucleus did not break down and one can speak of "rejuvenation" of the old nucleus by the surrounding young cytoplasm.

As we shall now see, although a giant green alga seems very different from an amphibian or sea urchin egg, there are remarkable similarities in the nucleocytoplasmic interactions in all cases. We are again faced with the paradox if unity and diversity in living organisms.

BIOCHEMISTRY OF ANUCLEATE FRAGMENTS OF EGGS

Amphibian Eggs

The huge germinal vesicle plays only a passive role in the biochemical changes (increase in protein synthesis with preferential histone synthesis, stimulation of protein phosphorylation, production of the maturation promoting factor) which take place when maturation is induced in fullgrown *Xenopus* oocytes. All these changes occur even if the oocytes have been enucleated prior to progesterone addition. It has also been recently shown that a "heat shock" (heating for a few minutes above 30 °C) induces the synthesis of a few new proteins in *Xenopus* oocytes. These heat shock proteins are the same whether the oocyte possesses its nucleus or not. Enucleated *Xenopus* oocytes contain the cytoplasmic factors required for "reprogramming" injected nuclei. The cytoplasm of these oocytes contains factors which inactivate certain genes and reactivate others. The outcome of the experiments in which protein or 5 S RNA synthesis has been followed after injection of adult nuclei is the same whether the oocytes have a nucleus or not. Finally, enucleated oocytes undergo the same permeability changes and the same cortical reaction after parthogenetic activation as intact progesterone-treated oocytes.

However, we have seen that the mixing of the nuclear sap and the cytoplasm is required for swelling of injected nuclei, condensation of chromatin and aster formation. It has been reported that enucleated amphibian oocytes undergo repeated cleavages after injection of sea urchin mitotic apparatuses: however, it is possible that in such experiments, sea urchin chromosomes take part in the cleavages.

Cleavage in the absence of the nucleus has been obtained repeatedly in amphibian eggs. In these experiments, the eggs were fertilized and the egg and sperm nuclei were destroyed by pricking or irradiation. In general, partial blastulae, in which only the animal pole has cleaved, are obtained; gastrulation never takes place. The morphogenetic potencies of these *"achromosomal"* blastulae have been tested by grafting nonnucleated "cells" from arrested blastulae onto inducing sites of normal gastrulae. The grafts survived up to 4 days, but showed no sign of differentiation: nonnucleate ectoderm cells are thus totally unable to respond to the inducing stimulus of a normal organizer. These experiments show that the morphogenetic potencies of enucleated amphibian eggs are much lower than those of anucleate fragments of *Acetabularia:* they are limited to the capacity to divide repeatedly. We shall now see that the same is true for anucleate fragments of sea urchin eggs, which have been better studied from the biochemical viewpoint.

Sea Urchin Eggs

Biological Observations

The only materials that can be obtained in sufficient quantities for easy biochemical analysis are the anucleate fragments of sea urchin eggs. The unfertilized eggs of *A rbacia* can be cut into two halves by centrifugation in a density gradient. The heavy red pigment accumulates to the centripetal end of the egg, which successively elongates, takes a dumbbell shape and separates into two halves. The light, transparent half contains the nucleus, fat droplets, mitochondria and a clear, ribosome-rich layer, the hyaloplasm. The red heavy half is anucleate; it contains the red pigment granules and most of the yolk. Electron microscopy shows that the light halves also possess some of the yolk platelets and that mitochondria, as well as lipid droplets, are present in both halves. All kinds of egg inclusions are therefore presen in the two halves, but they are in different proportions; the only specific cell organelles are the nucleus, which is always in the light half, and the pigment granules, which are found only in the heavy halves.

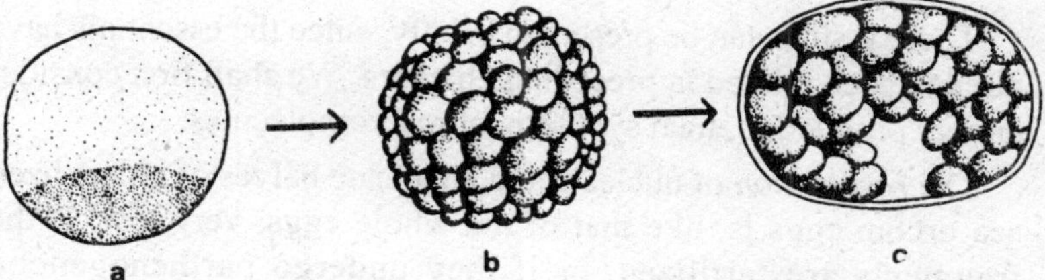

Figure 8.3 : a Anucleate half of a sea urchin egg. b Anucleate morula after 13 h. c Anucleate blastula after 3 days.

Both kinds of fragments can be fertilized or parthenogenetically activated by the classical method of Loeb (hypertonic seawater and butyric acid treatments). Nucleate halves develop very well after *fertilization* and give rise to normal plutei. The development of fertilized anucleate halves is less normal and plutei are not easy to obtain. This rather poor development is probably not so much due to haploidy as to the excess of yolk, which makes cleavage difficult. *Parthenogenetic* development of the nucleate halves is almost as good as that of normal eggs. But the development of the anucleate halves, after parthenogenetic treatments, is completely abortive. Asters can repeatedly form in the cytoplasm of the activated anucleate halves and irregular furrows can cleave the egg to form a kind of morula. But the typical cleavage pattern of the sea urchin eggs, with micromeres, mesomeres and macromeres is always missing. The blastula-like larvae neither hatch nor form cilia. As one can see, there is no true morphogenesis and we are very far indeed from the formation of the elegant, species-specific cap of *A cetabularia.* Only the formation of asters and furrows displays some autonomy toward the nucleus in activated anucleate sea urchin eggs.

The only other anucleate materials that have been obtained from eggs and have been the subject of rather limited biochemical studies are the *polar lobes,* which can be separated at the "trefoil" stage from the eggs of the mollusks *Mytilus* and *Ilyanassa.* They never form asters and do not cleave; all one can see are localized contractions of the cortex, which confirm that contractility of the cortical proteins displays some autonomy toward the nucleus.

Biochemical Studies

These results can be presented briefly, since the essentials have already been related in preceding chapters. We shall first consider energy production, then synthesis of macromolecules.

The *respiration* of nucleate and anucleate halves of unfertilized sea urchin eggs is, like that of the whole eggs, very low. If the fragments are fertilized, or if they undergo parthenogenetic activation, their oxygen consumption quickly increases. The respiration of the anucleate halves becomes even higher than that of the nucleate ones. This proves that the part played by the female pronucleus, in cellular oxidations, is negligible. This conclusions agrees with the already mentioned finding that the oxygen consumptionn of the isolated germinal vesicle of frog oocytes is hardly measurable. In fact, enucleated and normal frog oocytes have about the same respiratory rate. The higher oxygen consumption of the anucleate halves is probably related to the fact that, as shown by electron microscopy, the nucleate and anucleate halves contain, after their separation by centrifugation, two different populations of mitochondria. They differ in size, structure and probably in respiratory enzyme content. In fact, the anucleate, heavy halves contain a larger proportion of the reducing mitochondrial respiratory enzymes (dehydrogenases) than their nucleate counterparts.

Isolated polar lobes of the mollusk *Mytilus* have a low respiratory rate, which does not increase with time, in contrast with the blastomeres where the nuclei multiply at a fast pace.

Coming now to the synthesis of macromolecules, we have already seen that *protein synthesis is* independent of the presence of the nucleus in sea urchin eggs. Activation of anucleate fragments is followed by a strong stimulation of protein synthesis and is even stronger in activated anucleate halves than in nucleate ones. This stimulation is accompanied by an increase in the number of polysomes in anucleate as well as in nucleate halves; these polysomes result from the binding of "unmasked" maternal mRNA molecules to the pre-existing ribosomes.

A whole spectrum of proteins are synthesized after activation in both nucleate and anucleate halves, but it is not yet known whether the neosynthesized polypeptides are identical or slightly different in the two halves. Among the stored mRNAs, which are translated by both kinds of activated fragments, those coding for *histones* and for *tubulin* deserve special interest for the following reasons. In anucleate halves, there is of course no replication of nuclear DNA and no assembly of DNA in nucleosomes; yet histones are synthesized, apparently uselessly. This definitely proves that there is not necessarily a close link between histone production and nuclear DNA synthesis. Synthesis of tubulin in activated anucleate halves also raises an interesting question: Is translation of tubulin mRNA involved in the formation of asters in activated anucleate fragments? It is well known that unfertilized sea urchin eggs have a large pool of unpolymerized tubulin molecules. The formation of asters and spindle, in the fertilized sea urchin eggs, results from their polymerization around microtubule organizing centers (MTOC), such as the kinetochores of the chromosomes and the centrioles. If anucleate halves are treated with heavy water

(D_2O), many cytasters appear, which are centered around small centrioles or precentrioles which have apparently been assembled de novo. It is not known whether translation of tubulin mRNA is involved in the formation of asters in activated anucleate sea urchin egg fragments. But it is a striking fact that, despite tubulin synthesis, anucleate activated halves never form cilia; anucleate amphibian blastulae also lack cilia. Something necessary for the assembly of cilia is thus missing in nonnucleate cytoplasm. Whatever this might be, it seems clear that its production, in blastulae, requires transcription of nuclear genes. Anucleate parthenogenetically activated sea urchin fragments are also unable to hatch: they apparently cannot synthesize the "hatching enzyme". It would be interesting to know whether they possess a maternal mRNA encoding this proteolytic enzyme.

The *distribution* of the maternal mRNAs in the nucleate and anucleate halves of sea urchin eggs (obtained by high speed centrifugation) has been recently studied. It was found that total RNA, the bulk of the poly A containing mRNAs, actin and tubulin mRNAs are accumulated in the heavy, anucleate halves. This might explain why protein synthesis is stronger, after parthenogenetic stimulation, in anucleate than in nucleate halves. In contrast, histone mRNA is accumulated in the nucleate halves (even in uncent-rifuged eggs cut into halves with a glass needle). However, isolated nuclei contain little histone mRNA. The distribution of the various mRNAs is thus heterogeneous in sea urchin eggs, which is probably due to the organization of the cytoplasm in gradients.

If we compare amphibian and sea urchin eggs with *Acetab-ularia,* a striking fact emerges: although all anucleate fragments possess a large store of mRNAs, only the alga is capable of true morphogenesis. The reasons why, in eggs, the maternal mRNA store (theoretically capable, as we have seen, to synthesize more than 10,000 different proteins) does not allow even such early events as hatching and ciliation, are not known. It might be that the mRNAs are less stable in anucleate than in nucleate fragments. However, it is known that polyadenylation of the same mRNAs occurs in both kinds of fragments and that the enzyme respon-sible for polyadenylation is accumulated in the heavy anucleate halves. As we know, one of the functions of poly A addition is to increase mRNA stability in the cytoplasm. Transcription of nuclear genes during and after cleavage is thus an absolute prerequisite for further development. In molecular terms, the maternal mRNA store corresponds to *preformation,* the neosynthesized mRNAs to *epigenesis:* that development is an epigenetic process is, of course no surprise for embryologists. Development stops, in enucleated eggs, at an early cleavage stage probably because *they* are unable to synthesize the mRNAs and the specific proteins which are required for further developmental events.

Both nucleate and anucleate fragments of sea urchin eggs possess a large population of *mitochondria.* Since they contain DNA, the DNA content of nucleate and anucleate halves is about the same (the haploid female pronucleus represents only a small proportion of total DNA). After activation, both kinds of fragments synthesize small amounts of mitochondrial RNAs (in particular, mitochondrial ribosomal RNAs). The mitochondria of activated anucleate fragments synthesize as RNA species which can diffuse into the cytoplasm, but, since it is unable to bind to the cytoplasmic ribosomes, it cannot play an important role in the strong stimulation of protein synthesis which follows activation of anucleate halves.

An unexpected fact has recently come to light: *mitochondrial* DNA, RNA and protein syntheses are more strongly stimulated by parthenogenetic activation in anucleate than in nucleate halves of sea urchin eggs. It seems that removal of the nucleus induces a multiplication of the mitochondria, as well as an increase in mitochondrial DNA transcription and translation. If so, the often neglected female pronucleus would exert a *negative* control on the mitochondrial genome. This is reminiscent of the negative control exerted by the *Acetabularia* nucleus on the chloroplastic genome. In this alga, removal of the nucleus speeds up DNA, RNA and protein synthesis and we have seen that the chloroplasts are largely responsible for this increase in macromolecule synthesis.

Removal of the *polar lobe* at the trefoil stage results, in the mollusk *Ilyanassa,* in developmental abnormalities. Isolated polar lobes synthesize many proteins and thus contain, like sea urchin eggs, a store of maternal mRNAs. Removal of the polar lobe does not affect RNA and protein synthesis before relatively late stages of embryogenesis: no changes can be detected before the first appearance of developmental abnormalities. There is no evidence so far that the polar lobe synthesizes specific proteins and we have therefore no indication for a selective localization of certain translatable mRNAs in this region of the *Ilyanassa* egg. Curiously, the pattern of protein synthesis changes when isolated polar lobes are allowed to age. These changes are the same as those observed in intact embryos and should result from a selective translation of the maternal mRNAs (which is responsible for 98% of protein synthesis during early embryogenesis in *Ilyanassa)*. A recent study concludes that the cytoplasm of the polar lobe contains both repressors of translation and activators for unmasking maternal mRNAs: if so, both positive and negative controls of mRNA translation would operate in pure, nonnucleate cytoplasm.

THE IMPORTANCE OF THE NUCLEUS FOR EMBRYONIC DEVELOPMENT

General Background

Overwhelming evidence demonstrates that integrity of the nucleus is required for development beyond the blastula stage: haploidy, aneuploidy (unequal distribution of the genetic material in the blastomeres), in some cases hybridization. Many mutations are sooner or later *lethal* for the embryo. We shall concentrate on the cases in which lethality is an early event (arrest at gastrulation or neurulation) because they demonstrate that even primary morphogenesis is under nuclear control.

Haploidy can be obtained, in amphibian eggs, by a variety of means (pricking with a needle, fertilization with UV-irradiated sperm, removal of the maturation spindle just after fertilization). Whether the haploid embryos arise from the egg nucleus *(gynogenesis)* or the sperm nucleus *(androgenesis)*, the result is the same. After normal cleavage, gastrulation is slowed down and further development leads to the already mentioned *haploid syndrome,* which is characterized by microcephaly, deficiency in blood circulation, reduction of the gills, oedema, ascites and finally death at an early larval stage. Lethality seems to be due to haploidy per se, and not to lethal genes which would not be balanced by their normal alleles in the haploid condition. If, as may happen accidentally, the number of chromosomes doubles in a parthenogenetic egg before it undergoes cleavage, homozygous diploids are produced. Although they contain two copies of the lethal genes, they can reach the adult stage. This is apparently not true for the mouse in which haploid eggs fail to develop beyond the end of cleavage and parthenogenetic diploids are lethal.

In *polyspermic* eggs, only one of the nuclei of the supernumerary spermatozoa fuses with the egg pronucleus; the other nuclei remain haploid and may take part in the development. The resulting embryos are a mosaic of diploid and haploid nuclei; although part of the embryo is diploid, death occurs still earlier than in haploids. In mammals, dispermy, which takes place spontaneously in many species, including man, gives rise to *triploid* embryos. Triploids may also arise from fertilizationn of a diploid egg. It is lethal during the postimplantation development; only very few triploid embryos have been reported to have developed until term. Spontaneously and cytochalasin B-induced tetraploid mammalian embryos display the same lethality. Interestingly, mosaic embryos constructed by aggregation of diploid and triploid or tetraploid morulae are also lethal.

Aneuploidy (unbalanced chromosome complement) always has serious consequences: if, only one or two chromosomes are lost or added to the normal complement, viability is greatly reduced and morphogenesis is abnormal. If chromosome unbalance is very marked, as a result of multipolar mitoses during early cleavage, for instance, development stops at the blastula stage. Strong aneuploidy is thus as lethal as the complete absence of a nucleus.

A great deal of interest has been recently devoted to aneuploidy in mammals because it is the major cause of prenatal death and abortion or of severe postnatal congenital defects such as the human Down's syndrome. Monosomic mouse embryos (39 chromosomes instead of 40) die before or at implantation, while trisomic (41 chromosomes) embryos survive until at least day 10 of pregnancy. It is possible to rescue cells from such aneuploids by aggregating aneuploid morulae with normal diploid morulae. This strongly suggests that the chromosome unbalance is not lethal to the cells, but is specifically responsible for the impairment of morphogenetic events.

In the following, we shall deal mainly with lethal *hybrids,* because they have been better studied from the biochemical viewpoint than haploids, aneuploids and polyspermic embryos.

Lethal Hybrids

Biological Observations

Various things can happen when eggs are fertilized with sperm from a different species. The spermatozoon, in many cases, does not even touch the surface of the egg and nothing happens at all. In other cases, the spermatozoon comes in contact with the egg membrane and elicits the activation reaction, but it does not penetrate into the egg and does not even act as a parthenogenetic agent. In the remaining cases, the spermatozoon penetrates into the egg and amphimixy follows. But, in many instances, the paternal chromosomes, which have been introduced into the egg by the spermatozoon, are eliminated into the cytoplasm during cleavage; the eliminated chromosomes soon degenerate and development is of the haploid, gynogenetic type *(pseudohybrids).* In true hybrids, the paternal chromosomes are not eliminated; development is normal, for a longer or shorter time; then the hybrid embryos develop abnormalities and ultimately die. They are the so-called *lethal hybrids.* Finally, when the chromosomes of the two parental species are sufficiently similar from the genetic viewpoint, the hybrids are *viable;* they reach the adult stage, but they are not always able to reproduce. Sterility is frequent, because the homologous chromosomes cannot undergo perfect pairing during meiosis.

Early lethal hybrids will be the main subject of this chapter, because they have a peculiar interest for molecular embryology. When eggs of the common species of sea urchins or frogs are fertilized with sperm from other common species, development is often arrested at the late blastula or *early gastrula* stage. This is the stage at which, as we have seen, major biochemical changes occur in the embryo; the most conspicuous of them is the resumption of rRNA synthesis. Why the hybrid embryos are unable to go through this critical stage of gastrulation is the question that a number of biochemical studies have tried to answer.

Before we discuss the biochemical studies on early lethal embryos, a last important biological fact must be mentioned. If a piece of an amphibian lethal hybrid, which is arrested in development at the early gastrula stage, is grafted into a normal host, a "revitalization" usually follows. If, for instance, the dorsal lip of a blocked hybrid gastrula is grafted into a normal recipient gastrula, the grafted dorsal lip differentiates into chorda and induces, in the host, a secondary nervous system. This revitalization is not species-specific. The same result (chorda differentiation in the graft and neural induction in the host) is obtained when a dorsal lip taken from an hybrid between two species of frogs is grafted in a urodele (triton, axolotl) gastrula.

Revitalization has been explained by supposing that unspecific substances, which cannot be synthesized by the hybrid cells, are given by the host cells to the graft. But it has recently been shown that for certain lethal hybrids at least, it is sufficient to dissociate the cells of the blocked embryo and to cultivate them in salt solution to obtain extensive differentiation of the hybrid cells. The "conditioned media" prepared from such hybrid cells are very toxic for normal embryos. Therefore, it is possible that lethality is due, at least in part, to the production and retention of toxic substances (repressors) produced by metabolism.

Nuclear transplantation has been used in order to find out whether the block at gastrulation in the combination *R. pipiens y* × *R. catesbeiana d is* due to irreversible changes in the nuclei. Transfer of nuclei taken from arrested hybrid gastrulae into recipient enucleated *R. pipiens* unfertilized eggs showed that nuclei taken from hybrid embryos that have just stopped developing allow normal cleavage; but development stops abruptly at gastrulation (as in the hybrids themselves). If 'nuclei are taken 10 to 20 h after gastrulation arrest and then transplanted into enucleated unfertilized eggs, cleavage is no longer normal. There are thus no irreversible changes in the nuclei of the lethal hybrids so long as they do not undergo cytologically visible alterations.

Biochemical Studies on Lethal Hybrids

Respiration. Most of the biochemical studies made on lethal hybrids between sea urchins have been done with the combination *Paracentrotus* × *Arbacia.* These hybrids cleave normally, hatch and die during gastrulation.

Paracentrotus (P) eggs have a higher oxygen consumption than those of *Arbacia (A).* The respiration of the P × A hybrid is first similar to that of the maternal species (P). Later on, the respiratory rate decreases in the hybrids, so that intermediary values between those typical for P and A are obtained. This reduction of the respiratory rate in the P × A hybrids suggests that a negative control, due to the paternal A genes, becomes effective a few hours before the arrest of development.

More data are available for hybrids among amphibians. The best studied combination is one among two American species of frogs: *Rana pipiensy x Rana sylvaticad,* where development stops at gastrulation. Respiration shows a normal increase during cleavage, then remains absolutely constant until the death of the embryo several days later. This decrease in the respiratory rate affects all parts of the embryo, as shown by measurements of the oxygen consumption made on dissected hybrid gastrulae. The decrease in respiration rate is the same in the ectoblast, the mesoblast and the entoblast. The reduced respiratory rate of the hybrids is linked, as one would expect, to a diminution of carbohydrate metabolism. Glycogenolysis is slowed down, glycolysis is decreased and the ATP content drops more quickly under anaerobic conditions in the hybrid gastrulae than in the controls after development has stopped.

These results cannot be generalized to all hybrid combinations between amphibians, however. The rule seems to be that as in sea urchin hybrids the oxygen consumption still increases, but at a slower rate than in the controls in arrested embryos.

Nucleic Acid and Protein Synthesis. We shall first consider the lethal hybrids of echinoderms, then those of amphibians.

Protein synthesis has been studied in several combinations of *sea urchins.* The "pool" of soluble amino acids and small peptides remains of the maternal type, even at a time when an effect of the paternal nucleus should become evident. In the *Paracentrotusy* × *Arbaciaa* combination (P × A), paternal proteins (antigens) can be detected, by immunological methods, in the hybrid blastula. The amount of paternal protein synthesized is about half that which would have been produced by a diploid *Arbacia (AA);* it is identical with the quantity of *Arbacia* antigen which is produced by a haploid, parthenogenetic *Arbacia* embryo at the same stage. These antigens are not synthesized at all by anucleate fragments of *Arbacia.* It follows that the amount of antigen produced depends on the number of *Arbacia* genes present. It does not matter, in this respect, whether a haploid set of genes is placed in its own *Arbacia* cytoplasm or in the foreign *Paracentrotus* cytoplasm.

However, in other echinoderm combinations *(Strongylocentrotusy* × *Dendrasterd)* no paternal antigens can be detected in the hybrid embryos, which stop development at a later stage than in the P × A combination. In these hybrids, one enzyme has been characterized: it is the "hatching enzyme" which enables the blastula to dissolve the fertilization membrane and to swim freely in the sea. This hatching enzyme has been found in the hybrids to be of the purely maternal type. Taken together, these results suggest that in contrast to the P × A combination, the paternal chromosomes are partly or entirely inactivated in the hybrids which show a complete maternal dominance.

This suggestion has been reinforced by recent work on sea urchin hybrids which can reach the pluteus larval stage. Among many proteins synthesized by these hybrids, only a few are specified by the paternal genome. Analysis of mRNA production in these hybrids has further shown that the paternal mRNAs are strongly underrepresented; however, paternal histones and their mRNAs are detectable already during early cleavage. It has also been found that some of the messages coding for paternal proteins are not translated in the hybrids. Production of foreign paternal proteins in the hybrids is thus repressed at two levels: transcription of the paternal genes and translation of the mRNAs (except for the histone genes and their mRNAs).

The controls exerted by the egg cytoplasm on the activity of the paternal genes differ whether development of the hybrids stops at gastrulation or reaches the pluteus stage. A thorough study, with the now available methods, of the early lethal hybrids is needed before one can explain this difference. The present evidence suggests that arrest at gastrulation takes place in eggs in which the foreign paternal genes are not repressed by the egg cytoplasm and are thus expressed. Development until a more advanced larval stage would be possible when the paternal genes are repressed to a large extent during early development. In this case, development would be almost parthenogenetic as in gynogenetic pseudohybrids. However, in these hybrids the foreign paternal chromatin far from being eliminated, is repeatedly replicated.

Arrest of DNA synthesis is certainly *not* responsible for the block in development. DNA synthesis continues when the P × A embryos are blocked at the gastrula stage, but the rate of synthesis is slower than in the controls.

This probably results, from a discrete elimination of paternal chromosomes during the development of the P × A embryos. Such an elimination also explains why the blocked P × A hybrids contain three times more maternal than paternal DNA, as shown by molecular hybridization.

The situation in *frog hybrids* is not fundamentally different, but a few peculiarities should first be mentioned. Hybridization modifies the properties of the plasma membrane; its permeability to amino acids and phosphate increases. Furthermore, if cells of a lethal hybrid gastrula are dissociated by removal of calcium and magnesium from the medium, they have great difficulties in re-aggregating when these ions are added back to the medium.

These differences in reaggregation in hybrid and normal gastrula cells are probably due to differences in the surface glycoproteins; it is likely that changes in cell surface glycoproteins and cell adherence play an important role in the arrest of many hybrid combinations at gastrulation. Indeed, it is known that treatment of normal *Xenopus* morulae with lectins (which bind to the sugar residues of the surface glycoproteins) prevents their gastrulation.

Another characteristic of the lethal hybrids among frogs is that the *yolk re*serves are utilized at a much slower rate than in the controls. The enzymes present in the yolk platelets also fail to be synthesized or liberated at the normal rate. The reasons for this delay in vitellolysis are not known.

As in the lethal sea urchin hybrids, *paternal antigens* become detectable at the blastula stage and an additional basic protein is found in the hybrids.

The synthesis of several *enzymes* has been studied. Esterases remain of the maternal type in one lethal combination of frogs. In another, which is viable, the enzymes that catalyze the oxidative metabolism of the sugars have values that are intermediary between those found in the two parental species, but if the egg nucleus is removed before fertilization, in order to obtain *a merogone hybrid* (which is lethal at a later stage than gastrulation) the synthesis of these enzymes follows the paternal pattern.

As in the sea urchin hybrids, *DNA synthesis* does not stop, but only slows down after the arrest of devṣlopment. The same is true for historic synthesis, which remains well coordinated with that of **DNA.** However, DNA and **RNA** syntheses stop before protein synthesis. When the hybrid is close to death and can no longer synthesize nucleic acids, it becomes comparable to the anucleate fragment of an egg, in which protein synthesis can continue for a long time because of the preformed maternal **mRNAs** and ribosomes.

A situation somewhat different from lethal hybridization has been achieved by Gurdon who injected, into an enucleated unfertilized egg of *Xenopus,* a diploid nucleus taken from a blastula of *Discoglossus*. In this case and in contrast with the lethal hybrids, a diploid foreign nucleus, but no maternal haploid nucleus, is present in the egg. The eggs that have received such a foreign *Discoglossus* nucleus stop development, as many lethal hybrids do, at the late blastula stage. **RNA** synthesis has been analyzed in these blocked blastulae. There is no **rRNA** synthesis and there is a decrease in **mRNA** and tRNA synthesis.

As one can see, much more work is needed before we can understand the molecular reasons for the lethality in both sea urchin and amphibian hybrids. If one can venture a guess, arrest of development, followed by cell death, might largely be the result of the accumulation of foreign, paternal proteins.

BIOCHEMISTRY OF EARLY DEVELOPMENTAL MUTANTS

There is no doubt that molecular embryology would make tremendous progress if *mutants* deficient in a specific metabolic step were easily available. It is needless to recall that molecular biology would not be what it is today without the extensive study of thousands of mutants in bacteria and phages.

We have a few mutants available in amphibians and their number is increasing every year, but extremely few of them have been submitted to a biochemical analysis. This is unfortunate, since the already discussed *nucleolusless* (nu-o) mutant of *Xenopus* has played an exceedingly useful, if not decisive, role in the understanding of the intervention of the nucleolar organizers in **rRNA** synthesis and ribosome formation.

The only mutants that have been, to a very limited extent, analyzed from the biochemical viewpoint are the o and *cl* mutants of the axolotl. These two mutations lead to early lethality when they are in the homozygous state. o denotes *ova deficient* and *cl* denotes *cleavage arrest.* In the heterozygous condition, the o mutants reach the adult stage and are fertile, but many of the individuals die before they become adult. The homozygous o/o mutants die before or during gastrulation. Interestingly, they can be saved, as Briggs discovered, if one injects into the just fertilized egg the content of a germinal vesicle taken from a normal oocyte. The injection of this nuclear sap into one of the two first blastomeres enables the injected blastomere to develop further than gastrulation, while the other is blocked. Thus, the nuclear sap of the normal oocytes contains a factor, presumably a protein, that cannot be synthesized by the *o mutant* and that is necessary for development beyond the gastrula stage. Further studies on the chemical nature of this "corrective" factor will be awaited with great interest. We only know that it is present in the oocyte nuclear sap of a variety of amphibians and, thus, has no species specificity. The corrective factor is absent in the cytoplasm of ovarian oocytes, but is present in the germinal vesicle at all stages of *oogenesis*. RNA synthesis is very low in o mutants and it is, therefore, possible that the missing gene product is a karyophilic protein involved in RNA synthesis. Nuclear transplantation experiments have shown that nuclei taken from of o-morulae support full development after transfer into enucleated, unfertilized eggs; but nuclei taken from o/o blastulae no longer support it. Thus, irreversible alterations occur in the nuclei between the morula and the blastula stages in o/o mutants.

The *nc* (no cleavage) mutant of the axolotl is also of interest for both molecular cytologists and embryologists: the eggs of this mutant can be fertilized, but they do not cleave because they are unable to assemble their large store of unpolymerized tubulin into spindle and aster microtubules. The mutation can be corrected, and repeated cleavages will occur after injection of centrioles (basal bodies of flagella) into the cytoplasm of fertilized *nc* eggs. What seems to be missing in these eggs is the capacity to assemble microtubule organizing centers (MTOC).

Comparable results have been found in *Drosophila*, in which a mutation called *deep orange (dor)* causes early death in the homozygous *(dor/dor)* embryos; the defect can be corrected by injecting cytoplasm from unfertilized normal eggs into *dor/dor* blastulae. It is not yet known whether, as in the o mutant of the axolotl, the repair factor is accumulated in the germinal vesicle during oogenesis. Other developmental mutations of *Drosophila* have been corrected by injection of cytoplasm from normal eggs.

It has been known for many years that the complete absence of the X chromosomes *(nullo-x* condition) results in deficiencies in early *Drosophila* development: gastrulation soon becomes abnormal and then stops. Small deletions in the X chromosomes may also lead to early arrest of development. It is now well established that the blocks in early development mentioned in this section are due to the fact that mutations, such as o, *cl, dor, nullo-x,* etc. affect the organization of the *cytoplasm* during *oogenesis:* the normal alleles of these genes control the organization (in gradients, for instance) and the chemical composition of the egg cytoplasm. Unfortunately, the products of these very important genes have not yet been identified.

Mention should be made of a few mutations affecting mouse development. The Os (oligosyndactylia) mutation, located on chromosome 8, causes digit fusion in the feet of heterozygous mice. In the homozygous state, it causes lethality between the seventh division and implantation, as a result of a mitotic block in metaphase. However, unlike the *nc* mutation in the axolotl, the Os mutation does not impair tubulin synthesis or microtubule assembly. It seems to be associated with a defect in spindle formation after the sixth division. Another dominant mutation, located on chromosome 2 [*AY* (yellow)] is responsible for the yellow coat of the mutant mice; it also causes early lethality in the homozygous state. Cleavage is arrested before hatching from the zona pellucida and differentiation of the trophectoderm into trophoblast giant cells is totally impaired. Nothing is known about the biochemical target of the mutation.

The *t-complex,* located on chromosome 17, where many different mutant alleles are known, includes the H-2 locus which controls, in the adult mouse, the synthesis of the histocompatibility antigens responsible for graft rejection. Of particular interest, among the many t-mutants, are the t^{12}/t^{12} mutants where development stops already at the morula stage. Neither transcription nor translation are affected by this mutation until the embryo dies, but energy production must be somehow affected since neutral lipids accumulate in nuclear and cytoplasmic droplets.

"TRANSGENIC" MICE AND TERATOCARCINOMA

An old dream for many biologists has been to create new strains of animals with predetermined genetic changes: in other words, to introduce a selected gene into an egg and to witness its expression not only in the adult, but even in its offspring.

As we have seen, pure genes are now available in relatively large amounts due to progress in molecular biology. Parallel progress in embryology allows skilled experimenters to inject substances into one of the pronuclei of a fertilized mouse egg, to culture in vitro the injected eggs and to implant them into pseudo-pregnant females. If all goes well, fetuses will develop in their foster mothers and grow into adults, which will give birth to offspring after mating.

Injection of various genes into mouse pronuclei has been successfully achieved in recent years: if the gene coding for rabbit /3-globin, for instance, has been injected into one of the pronuclei of a fertilized mouse egg, adult mice, in which all the cells contain several copies of the foreign (rabbit) gene, can be obtained. Still more important, these mice often transmit the gene to their offspring (or some of their offspring). These so-called transgenic mice fulfill the old dream of directing mutations, of manipulating Heredity at will. Success requires the integration of the foreign gene in the germ line cells of the recipient eggs and embryos.

Brinster has recently succeeded in performing similar experiments with "fusion-genes". A fusion-gene has been made of the promotor/regulator of the metallothionein I gene from the mouse and the structural human growth hormone gene. In mammals, metallothionein is normally produced by the liver and growth hormone by the anterior pituitary. When this fusion-gene is microinjected, the liver of the transgenic mice produces large amounts of human growth hormone, whereas their anterior pituitary does not. As shown in Fig. 68, such mice become giant mice. The same experiments are now feasible with eggs of domestic animals and could possibly lead to the economically important production of giant cows and pigs. From a more fundamental viewpoint, the most important feature of these experiments is that they offer a new way of studying gene regulation during mammalian development. It is now possible to learn much about the *tissue-specific regulation of* the expression *of* foreign sequences coding for globin, growth hormone, insulin, transferrin, metallothionein, imminoglobins. etc.

It has been shown that insertion of foreign genes in the germ line cells can ce mutations affecting development. The first case of *insertional mulagenesis* rice was reported by Jaenisch: insertion of viral DNA sequences in mice red in a recessive lethal mutation, which causes, in the homozygous state, the death of the fetus at day 14 of pregnancy. Using the viral DNA as a probe, it was ible to show that the insertion lies in the 5' end of the al (I) collagen gene; *h* of the embryos could be due to their incapacity to synthesize collagen. Inonal mutations have also been reported in mice carrying human growth hore gene sequences; we have seen that insertion of a mobile genetic element into a *Drosophila* gene ınduces its mutation.

Similar experiments with human eggs have not yet been reported and it is obvious that very important ethical problems would arise if they were attempted. We all know that many "babies in test tubes" were born and that they grow normally. The methodology of artificial insemination, in vitro culture and replantation in a foster mother works as well for the human as for the mouse egg. "Transgenic men" are thus a theoretical possibility. Let us hope that if human eggs are ever manipulated (attempts to introduce new genes or to replace a "bad" gene by a "good" one), it will always be for the good and never for the evil of mankind. In this field, the moral responsibility of the scientist toward mankind is very heavy, not to say frightening.

Exciting material for both embryologists and oncologists is the *mouse teratocarcinoma,* a malignant tumor which originates spontaneously from the ovary or the testis, or can be induced by grafting 7-day-old embryos into the testis of male mice. Some teratocarcinoma cells differentiate into a monstrous embryo; others, called embryonal carcinoma cells, remain undifferentiated and are responsible for the malignancy of the tumor. If these "stem cells" are injected into a mouse, the recipient animal will be killed by a fast-growing tumor made of a variety of tissues as well as undifferentiated embryonal carcinoma cells.

The *pluripotentiality* of the embryonal carcinoma cells makes them attractive to embryologists because they share this property with the ectodermal cells of the inner cell mass. Particularly interesting is the fact that if a small number (three to five) of these cells are implanted into a normal mouse blastocyst, they are cured from their malignancy and can participate in normal development. Rccently, teratocarcinoma cells have been used as vectors for *gene transfer* in mice: human β-globin genes have been introduced into mouse embryonal carcinoma cells by growing them in the presence of cloned human globin DNA sequences. The transformed cells may be introduced into normal blastocysts which are then grown in the usual way. The expected result is the production, in the second generation, of transgenic mice, thus of new strains with predetermined genetic changes (production of human β-globin) by another method rather than direct injection of a gene into a pronucleus.

As one can see, embryology is entering into a new era: fascinating, exciting, perhaps even dangerous discoveries lie ahead of us.

CONCLUSIONS

It is now time to draw a few conclusions before we leave the early stages of development and move into the much more complex area of embryonic differentiation. The work that has been summarized in the preceding chapters clearly shows that the chemical machinery for the initial steps of development has already been made during oogenesis. Not only ribosomes have been produced in such a high number as to allow development until a late larval stage (feeding tadpole) without further synthesis of rRNA or ribosomal proteins, but a great variety of mRNAs, containing a large array of information, have been stored in unfertilized eggs. New mRNA species are synthesized during cleavage, but they are not actually needed during this period which can normally proceed in the complete absence of RNA synthesis (except in mammals). The result of this economic, almost avaricious policy is that during the early stages of development, the importance of the cytoplasm largely overwhelms that of the nucleus. The critical point is the end of cleavage; afterward, gastrulation movements, and neural induction become impossible without fresh information, resulting from gene activation and transcription.

But one should not forget that the fundamental, basic processes that control final development occur during these very early stages. Polarity, which will control cephalocaudal differentiation, formation of the grey crescent in amphibian eggs, (an immediate consequence of fertilization, responsible for the dorsoventral organization of the egg), germinal localizations in the "mosaic" eggs and embryonic regulation, all occur at stages in which the genes are almost entirely repressed and mRNA synthesis is hardly detectable.

If one returns to the different theories of embryonic differentiation, one cannot escape the conclusion that the ideas expressed by Morgan, almost 50 years ago, remain valid as far as the early stages of development are concerned. His concept that gene activity is controlled, at these early stages, by the properties of the heterogeneous cytoplasm which surrounds equipotential cleavage nuclei has been substant-iated by many experiments (nuclear transplantations, in particular). It is now clear that the nuclei of adult, differentiated cells and those of zygotes possess the same genes, but cytoplasmic factors present in oocytes and eggs affect their expression. The spectrum of proteins synthesized under the control of an adult nucleus is not the same whether it has been injected into an oocyte or it lies in its own cytoplasm. The identity of the *cytoplasmic factors* present in oocytes and eggs, which allow the reprogramming of injected nuclei, remains unknown. If one may venture a guess, these factors might well by karyophilic proteins, which have the remarkable capacity to move quickly from the cytoplasm into the nuclei where they soon accumulate and may bind to DNA. Detailed analysis of the proteins present in the oocyte nucleus and in cleaving egg nuclei is an important and urgent task for the molecular embryologists of tomorrow.

The cytoplasmic factors that have just been discussed certainly continue to play their role in later stages of development, when the cells undergo their typical differentiation from the morphological, physiological and molecular viewpoints. But their importance becomes less and less evident as development proceeds. The sequential, selective activation of individual genes or groups of genes now appears in the forefront.

9

How Cells Differentiate

GENERAL INTRODUCTION

The basic problems involved in embryonic differentiation have already been discussed at the beginning of this book. A typical question, which we have already asked before, is the following: How is it possible that red blood cells, which have synthesized large amounts of a very specific protein, hemoglobin, appear only in a very localized area of the embryo, at a very precise stage of its development? A good example of embryonic cell differentiation; it is a section through a young tadpole and shows, side by side, two types of highly specialized cells, which have just undergone embryonic differentiation. *Cartilage* cells and *muscle* cells lie side by side and are easily recognizable by their morphology. The cartilage cells are surrounded by an amorphous matrix made of special glycoproteins, called *chondroproteins;* the muscle cells contain fibrils (the myofibrils), which are made of entirely different proteins, in particular, the contractile proteins *myosin* and *actin.* The structure, the function and the biochemical composition of the cartilage and muscle cells are thus very different and specific for each type of cell. Yet, if the section had been made a couple of days earlier, it would not have shown such highly differentiated cells; histological studies of the embryonic development, during this short 2-day period, would have shown that cartilage and muscle cells have *the same* mesodermal origin. In a group of mesodermal cells, which all look alike, if not identical, some cells will form glycoproteins in large amounts and become cartilage; others will synthesize contractile proteins and differentiate into muscle cells.

Since we know, due to the nuclear transplantation experiments of Gurdon, that all the nuclei of the tadpole and even the adult contain the same genes as the fertilized egg (the zygote), differentiation can only be explained by assuming that in muscle cells, certain genes that direct the synthesis of the contractile proteins are active, while those that direct the hemoglobin and chondroproteins are silent. In the same way, the hemoglobin genes must be active in the differentiating red blood cells and the chondroprotein genes must be functioning in the young cartilage cells. Thus, the genes involved in the synthesis of large amounts of the other specific proteins (which Holtzer called "luxury" proteins) must be inactive in these cells. On the other hand, the genes that control the synthesis of proteins which are needed by all cells must necessarily always be active. Any cell, whether it is a muscle cell, a red blood cell, a cartilage cell, etc.; requires energy and must possess all the enzymes needed for oxidative phosphorylations (cytochrome oxidase, dehydrogenases, etc.). Differentiated cells contain variable amounts of these enzymes; Holtzer called them the "housekeeping" proteins, and they must be present if the cell is going to survive. Thus, the problem of cell differentiation is largely a problem of *gene regulation.* We have seen that many different mechanisms may operate in the control of gene activity. During cell differentiation, where one or a few genes coding for specific luxury proteins are selectively expressed, there is little doubt that the main control takes place at the *transcriptional* level. As we have seen, "active chromatin" displays several peculiarities: high sensitivity to DNase digestion, frequent undermethylation of the DNA molecule and quantitative or qualitative changes in the DNA-binding proteins are well documented to day. But this is still a crude approach and we do not yet know how specific genes are turned on or off in a very precise and delicate way. We do not even know by which mechanisms the hemoglobin genes are undermethylated in the chromatin of red blood cell precursors and fully methylated elsewhere. How apparently undifferentiated cells become "committed" to differentiate along a given developmental pathway (erythropoiesis, myogenesis) remains, despite many efforts, an unsolved problem.

One of the hypotheses proposed to explain cell differentiation is that of "quantal mitoses", cell divisions which would give rise to two *non-identical* cells: one of them would remain an undifferentiated *stem cell,* the other would be committed to differentiate according to a given program. By repeated cell divisions, this committed cell would ultimately give rise to *a clone* of nearly identical differentiated cells. This clonal theory of embryonic differentiation is now widely accepted. However, it should be pointed out that the analysis of allophenic (tetraparental) mice has shown that all the muscles of a mouse derive from a small number of cells committed to become myocytes (muscle cells) in the somites, i.e. they do not derive from a single clone of cells, but from a few independent clones of cells.

One usually speaks today of *terminal* differentiation, a term which seems to imply that cell differentiation is an irreversible process. As we shall see, there is at least one case in which differentiated cells may *dedifferentiate* and then redifferentiate along another pathway. A shift in the differentiation program, called *transdifferentiation,* is thus a possibility, but it remains a rare event in the animal kingdom.

It is often believed that there is an antagonism between *cell division* and *cell differentiation;* indeed, mitotic activity, in general, slows down or even stops completely when cells fully differentiate. But one should not be too dogmatic on this point. There are several examples in which dividing cells have been found to be fully differentiated, as evidenced by the production of specific luxury proteins.

Two different biological systems can be used for the study of cell differentiation: developing embryos and in vitro cultures of embryonic cells committed to differentiate along a particular developmental pathway. The two systems are adequate for the analysis of cell differentiation, but both have their advantages and disadvantages. Whole embryos represent a more "natural" system than cultured cells; although the latter are easier to work with, the results obtained with them do not necessarily apply to whole embryos. This is not surprising since, in embryos, in contrast to cultured cells, cellular movements, cell-to-cell interactions, morphogenetic gradients and inductions play a decisive role. When embryos are dissociated into individual cells, these cells lose the precise localization they had in the morphogenetic gradient where they interacted with their neighbors. The result is the loss of what has been called *positional information.* This, as well as the poor permeability to drugs of whole embryos from certain species (amphibians, in particular), explain why some agents which, as we shall see, suppress cell differentiation in cultured embryonic cells have very little effect on intact embryos.

It is impossible to give a full account, in a few pages, of all the work that has been done and is being done on cell differentiation. We shall concentrate on *embryonic* differentiation and speak more superficially of the studies made by the workers who use tissue cultures as their main method.

Before we come to the presentation and discussion of the known facts, a last word of introduction should be said. The study of cell differentiation has more than an academic interest. It is a problem which is of fundamental importance for mankind, because cancer is, to a large extent, a disease in which the cells divide, but fail to differentiate. The solution of the cell differentiation problem might lead sooner or later to a solution of the cancer problem. If malignant cells could be made to differentiate, they would stop dividing and would perhaps no longer be malignant.

SPECIFIC PROPERTIES OF CELL MEMBRANES

Already at the gastrula stage,-as-shown-by Holtfretex,, cells of the different parts of the embryo differ in their *tissue affinities.* If cells of the ectoblast and the entoblast are joined together in the same explant, they separate from each other after a few hours of culture. They have *negative affinities,* since they reject each other. On the other hand, the mesoblast cells stick to either ectoblast or entoblast when they are placed in contact; they have *positive affinities* for the cells of the two other cell layers. Interestingly, treatments with actinomycin or cycloheximide modify the ectoblast tissue affinities: their acquisition or maintenance thus requires the synthesis of macromolecules.

Comparable findings have been made with cells isolated from much older embryos. For instance, Moscona dissociated cells from the retina of chick and mouse embryos and mixed them with other dissociated cells, coming from the kidney of the same embryos. He found that retina cells recognize each other and form *a chimaeric* retina, made of mouse and chicken cells; in the same way, the dissociated kidney cells form *chimaeric* kidney tubules, made of cells originating from the two species of embryos. In this case, the cells recognize the other cells which belong to the same tissue and are members of the same family.

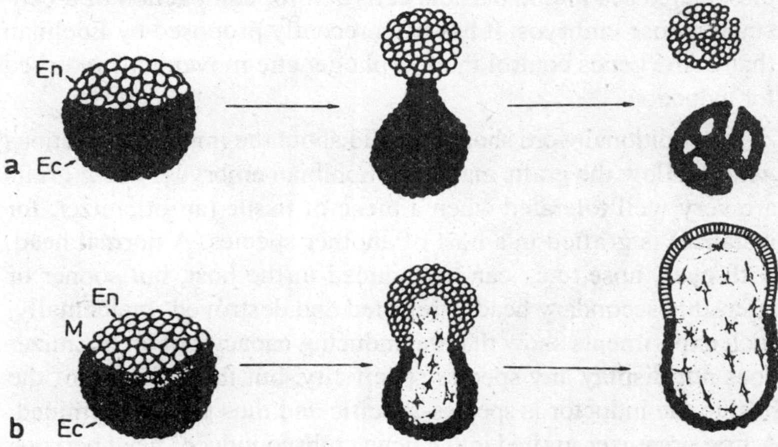

Figure 9.1 : a,b. Tissue affinities. a Ectoderm (Ec) and endoderm (En) show "negative" affinity; they quickly separate from each other and form unorganized gut and epidermic cells. b If a piece of mesoderm (M) is inserted between the two fragments, ectoderm and endoderm no longer separate; they both have a "positive" affinity for mesoderm; a vesicle containing mesenchymatous cells is obtained.

Many attempts are being made today to identify the molecules responsible for specific cell-to-cell adherence in embryonic tissues and in lower invertebrates, such as the sponges (where reaggregation of dissociated cells is easy to obtain). These molecules (called *CAM* for cell adherence molecules or Cognins) display tissue specificity: L-CAM from liver favors the reaggregation of liver, but not of nerve cells. In contract, N-CAM from the nerve cells acts specifically on the reaggregation of neurones and has no effects on liver cells. All these molecules are large acidic glycoproteins, with a molecular weight of more than 100,000. Schematically the relationships between N-CAM, a nerve cell and the outer medium. N-CAM is formed of two closely related polypeptide chains with molecular weights of respectively, 140,000 and 170,000. At the end of the amino terminal region is a binding site for another N-CAM molecule, while the other end (carboxyl terminal region) penetrates into the cell through the cell membrane. Specific cell adhesion molecules are already present at early stages of development and are responsible for the reaggregation of dissociated sea urchin blastula cells and for compaction of 8-cell-stage mouse embryos. It has been recently proposed by Edelman that CAM genes control the morphogenetic movements required for induction.

An additional word should be said about the immunity reactions which follow the grafts made in amphibian embryos. These grafts are very well tolerated when a piece of tissue (an organizer, for instance) is grafted in a host of another species. A normal head, with eyes, nose, etc., can be induced in the host, but sooner or later, this secondary head is rejected and destroyed. Incidentally, such experiments show that the inducing capacity of an organizer does not display any species specificity, but the response of the host to the inductor is species-specific and thus gene-determined. A frog organizer grafted into a newt embryo induces newt nervous tissue.

From the viewpoint of the immuno-response, there is a difference between amphibian and mammalian embryos. If a foreign tissue is grafted in the latter, it is not rejected even when the immuno-system has been built up after birth, (acquired tolerance).

EFFECTS OF TISSUE EXTRACTS ON CELL DIFFERENTIATION

Many efforts have been made in the past with the hope of isolating substances from differentiated tissues, which would induce the tissue-specific differentiation of undifferentiated cells. These efforts have so far been frustrating and the limited effects obtained should be ascribed to the presence of many factors in tissue extracts, which stimulate or retard cell proliferation and, as a result, may affect cell differentiation in an indirect way. Substances present in adult tissues which inhibit mitotic activity of homologous embryonic cells in a tissue-specific way are called *chalones*. For instance, an extract of adult kidney reduces the number of mitoses to a greater extent in embryonic kidney (pronephros) than in all other tissues of an amphibian larva. Many different chalones have been described; they seem to be very heterogeneous molecules, since their molecular weights may vary from less than 10,000 to more than 100,000.

It is beyond the scope of this chapter to discuss the numerous substances *(growth factors)* which stimulate cell division and may indirectly slow down differentiation. Many of them bear some similarities with the pancreatic hormone *insulin* and act by binding to specific receptors present on the cell membrane. The growth factor-receptor complex is internalized by endocytosis and partly broken down; the growth factors ultimately stimulate DNA replication in the cell nucleus. The best known of these growth factors is the *ectodermic growth factor* (EGF); this small polypeptide (6000 daltons, 53 amino acids) exerts many early effects on fibroblasts (increases in Na' uptake, intracellular pH, amino acid and glucose transport, etc.) which finally lead to the stimulation of nuclear DNA synthesis.

IMMUNOLOGICAL STUDIES ON EMBRYONIC DIFFERENTIATION

Immunological (serological) methods are exceedingly valuable for the detection *of specific proteins.* If a protein (an *antigen)* is injected into an animal, a rabbit for instance, this animal will produce in its serum an *antibody.* This is a protein which is called an *immunoglobulin* and which reacts in a highly specific way with the antigen. The antigen-antibody complex is less soluble than the antigen and the antibody. The presence of the antigen can thus be detected by its immunological precipitation after addition of the antibody. If one binds to the antibody, by chemical means, a fluorescent dye or a radioactive marker, it becomes possible to detect, under the microscope, the cells which contain the antigen. The present technology of *monoclonal antibodies* production makes antibodies of exquisite selectivity available; they are produced by *hybridomas,* which result from the fusion of a malignant myeloma cell and a cell taken from the spleen of an animal which has been immunized against the substance of interest.

Early work in which embryos crushed at various stages of their development were used as antigens showed that new antigens appear at amphibian gastrulation and that others can be detected at later stages of development. This merely confirms the data that physical methods for the separation of newly synthesized proteins had shown. It is probable that the use of monoclonal antibodies would lead further, but the most important application so far of immunological methods to embryology has been the analysis of the formation of the eye *lens* proteins (the α, β and γ-crystallins).

The *lens* is induced by the optic cup, at the expense of the ectoderm, at a late neurula stage. Immunological analysis shows that the lens antigens can be detected in the head of the young neurula. They are present in all cells, but in low concentrations. When the lens is induced, the specific lens antigens become detectable in the still undifferentiated lens in higher concentrations than elsewhere. Their concentration considerably increases at the time of cell differentiation, characterized by the appearance of fibers, made of crystallins, in the center of the embryonic lens. This is due to a selective activation of the crystallin genes under the influence of the inducing retina cells. There is thus a progressive restriction in the synthesis of the lens protein from the whole head to the eye lens only during development.

Another system of interest is the so-called *Wolffian regeneration of the lens* in adult amphibians or advanced tadpoles. When the fully differentiated lens is removed by surgical operation, a new lens forms at the expense of the superior rim of the iris. The various biochemical phases of lens regeneration have been very accurately studied by Yamada. The first event is *a dedifferentiation* of the iris cells, which lose their black pigment (melanin). Depigmentation is followed by DNA synthesis and cell multiplication. When mitotic activity stops, a period of intensive RNA synthesis sets in. Large nucleoli make their appearance and rRNA synthesis becomes very conspicuous. This phase of RNA synthesis is very important for the success of the regeneration. The latter is inhibited by actinomycin, which blocks RNA synthesis. The mRNAs for the *crystallins* are also synthesized during this period and these proteins become detectable, by immunological methods, immediately after the phase of extensive RNA synthesis. The various crystallins are synthesized one after the other. They are found, at first, in both the cytoplasm and the nucleus; the latter degenerate afterward and completely disappear. These experiments show that inducing agents originating from the retina produce a derepression of the genes present in the iris. As a result, synthesis of the crystallins is induced in the regenerating lens. Cell differentiation (formation of lens fibers) and synthesis of the crystallins are closely linked and go hand in hand. Wolffian regeneration of the lens is an interesting case of *transdetermination,* the phenomenon first discovered by Hadorn in transplanted imaginal disks of *Drosophila* larvae. A similar phenomenon, which has been called *transdifferentiation,* can be obtained with cultured neural retina embryonic cells: some of the pigmented cells lose their pigment and differentiate into lens cells. Small lenses, called lentoids make their appearance in the cultures. As in Wolffian regeneration, inactive crystallin genes are activated: crystallin mRNAs are produced in large amounts and the corresponding proteins accumulate in the lentoids

ENZYME SYNTHESIS AND EMBRYONIC DIFFERENTIATION

It has often been suggested that differentiation might be due to the induction of enzyme synthesis resulting from the appearance, in a region of the embryo, of the substrate for the enzymatic reaction, or to the repression of enzyme synthesis by the localized accumulation of the end product of the enzyme reaction. Such control mechanisms of enzyme synthesis operate efficiently in bacteria and the elucidation of their genetic and molecular bases is at the root of the famous JacobMonod model of gene regulation in bacteria. But eggs are not bacteria, and all efforts made to demonstrate substrate-induced synthesis in embryos have been unsuccessful.

The presence of many enzymes in unfertilized eggs makes an enzymological approach to the study of cell differentiation difficult: some of the enzymes synthesized during oogenesis are kept ready for use at much later stages of development. However, this approach has been valuable in a few cases, as we shall now see.

Several *hydrolytic enzymes* are known to increase in activity in the digestive tract when it becomes functional. This is true for amylase, proteases, phosphatases, etc. in sea urchin plutei as well as in amphibian tadpoles and chicken embryos. In such cases, enzyme synthesis is clearly linked to the acquisition of a physiological function.

The activity of *cholinesterase,* an enzyme which plays an essential role in the nervous system, is already important in the unfertilized amphibian eggs. It increases considerably when, after neural induction, the nerve cells begin to differentiate into typical neurones which are physiologically active. Similar changes in cholinesterase activity have been recorded in the cells of a tumor of the nervous system, *neuroblastoma.* The tumor cells have a low cholinesterase activity. but, if they change into more typical neurones (provided with nerve fiber expansions) as the result of a change in the in vitro culture conditions, cholinesterase activity very markedly increases as the result of the synthesis of new enzyme molecules.

Changes in enzyme synthesis, correlated with modifications of the outer medium, have been recorded in other systems as well. For instance, *glutamine synthetase,* an enzyme which is also involved in nervous system metabolism, is synthesized, at a given time of development, by the retina of the chick embryo. If the retina cells are dissociated, and then allowed to reaggregate, glutamine synthetase synthesis is accelerated by several hours. Still earlier synthesis of glutamine synthetase is obtained when the retina cells are treated by corticosteroid hormones. The use of macromolecule synthesis inhibitors suggests that glutamine synthetase synthesis is regulated at both the transcriptional and translational levels.

It is, of course, impossible to give here a complete summary of all the work that has been done on enzyme synthesis during development and we shall close the discussion with a last example, that is, the synthesis of the *pancreatic enzymes* (trypsin, chymotrypsin, amylase, etc.) which has been the subject of very interesting studies by Rutter. The pancreatic "acini," forming the glandular site of digestive enzyme synthesis, are induced by the surrounding *mesenchyme* (the still undifferentiated mesodermal tissue). Their induction is of the "transfilter" type: it is successful even if the reacting and the inducing tissues are separated by a millipore membrane. Rutter succeeded in isolating a protein from the filters, which certainly plays a great role in the induction. This protein induces in the reacting cells the synthesis of a particular kind of low molecular weight DNA. This synthesis is followed by the replication of the main DNA, which leads to sustained mitotic activity. Only when a sufficient number of cells becomes available does RNA and then protein synthesis occur. The various pancreatic enzymes are synthesized in a constant order, each following the other. There are no indications that enzymes which serve in an identical metabolic pathway, protein or sugar metabolism, for instance, are synthesized together. This speaks against the existence of an "operon," similar to those of the bacteria. It is clear that the structural genes coding for the different pancreatic enzymes are activated in a sequential and individual manner, since there is a concomitant increase in their respective mRNAs: for instance, the content of the pancreatic cells in amylase mRNA increases 400 times during their differentiation. In this case, the main control of enzyme production is clearly at the transcriptional level.

Transcriptional controls are particularly clear when organs respond to *hormonal* stimulation by the production of large amounts of a specific protein. For instance, steroid hormones induce the secretion of very large quantities of ovalbumin in the chick oviduct, and of vitellogenin in the amphibian liver. The hormone binds to a receptor in the target cell; in the nucleus, the hormone-receptor complex activates selectively the ovalbumin or the vitellogenin genes, probably by binding to them. As a result, chromatin in the structural gene and its flanking regions takes the loose conformation of "active" chromatin. The gene is now exposed to RNA polymerase activity and the final result is an enormous production of the corresponding mRNA, which is immediately translated into protein: the latter is finally excreted. While there is less than one molecule of ovalbumin mRNA in each cell of the unstimulated chick oviduct, thousands of molecules per cell are found after steroid hormone administration. As we have seen, the response of the amphibian oocyte to progesterone is fundamentally different since gene transcription has very little, if any, importance for protein synthesis. This process is controlled at the translational level.

DIFFERENTIATION OF CULTURED EMBRYONIC CELLS

A Brief Description of a Few Biological Systems

Numerous studies on the in vitro differentiation of undifferentiated, but already "committed" embryonic "blast" cells have led to the conclusion that the key event is the selective activation and expression of a small number of specific genes.

Myogenesis (muscle differentiation) is one of the favorite systems for the in vitro study of cell differentiation. Undifferentiated *myoblasts* fuse together in culture and form *myotubes,* whereby the latter finally differentiate into contractile fibers, and simultaneously, several muscle-specific proteins (in particular, muscle actin and myosin) make their appearance. DNA synthesis stops when myoblasts fuse together. Experiments with actinomycin have shown that myoblasts synthesize unstable mRNAs, while differentiated muscle cells contain stable mRNA molecules. Large amounts of the messengers coding for the muscle-specific proteins are produced when the myotubes differentiate into muscle cells. In contrast, the bulk of the mRNA population (about 17,000 different mRNAs) does not vary greatly during myogenesis.

Another interesting system is *chondrogenesis* (cartilage formation), in which embryonic *chondroblasts* differentiate into *chondrocytes* which incorporate inorganic sulfate and build up an extracellular matrix of sulfated proteoglycans.

Chondroitin sulfate and chondroproteins are typical markers of cartilage differentiation. Soluble factors released by in vitro growing chondrocytes stimulate differentiation of chondroblasts into cartilage.

There is growing interest in the differentiation of some fibroblast strains (preadipocytes) into *adipocytes* (fat cells). Treatment with insulin triggers or accelerates the accumulation of triglycerides in the cytoplasm which becomes filled with a large lipid droplet; simultaneously, the cytoskeleton breaks down.

Many papers have been devoted to *erythropoiesis,* the formation of red blood cells, which is characterized by the synthesis of an easily detectable protein, hemoglobin. Erythropoiesis is stimulated by the addition of *hematopoietin,* a protein hormone made mainly by the kidney. The molecular organization of the α- and β-globin genes coding for the two globin chains of hemoglobin is very well known: we can follow, at the gene level, the shift from embryonic to fetal and finally adult hemoglobin. We know that the α- and β-genes lie on different chromosomes, and that there is no amplification of these genes when large quantities of hemoglobin are synthesized during erythropoiesis. We also have a growing knowledge of the multiple molecular causes of the *thalassemias,* hereditary diseases in which there is an unbalance in the synthesis of the α- and β-globin chains. It is the dream of many molecular biologists to cure these diseases by transfer of the correct gene into the deficient genome of the patient (gene therapy); success is not very far ahead. Finally, a large amount of work is being done on Friend's murine *erythroleukemic* cells (MEL cells): these cells are malignant because they are infected by Friend's leukemia virus. They grow continuously in culture, but do not differentiate. However, differentiation, evidenced by the synthesis of hemoglobin and a few other marker molecules, can he induced by treating MEL cells with simple chemical agents (dimethylsulfoxide, butyrate, etc.). In all these cases, whether hemoglobin synthesis occurs spontaneously or is induced experimentally, there is a close parallelism between the number of globin mRNA molecules present in a cell and the amount of hemoglobin it synthesizes. Thus, during erythropoiesis, a small number of genes are activated sequentially.

Among the other systems in which cell differentiation can be followed in vitro, mention should be made of the formation of *melanocytes* (pigment cells) and *neurones* (nerve cells). Melanocytes contain a brown pigment, called melanin, which makes these cells easily recognizable under the microscope.

Neurones possess *neurites* (axones and dendrites), have typical electrophysiological responses when they are excited and contain a number of specific proteins. The outgrowth of neurites is stimulated by *a nerve growth factor* (NGF) which has been well characterized. Cell lines derived from brain tumors (neuroblastomas, neurogliomas) grow actively in culture; these cells are undifferentiated, but they can be induced to differentiate and to form neurites by treatments with a number of physical and chemical agents which tend to slow down their growth.

It has been possible to isolate, from teratocarcinomas, embryonal carcinoma cells (ECC) of different origins and to produce a large number of established cell lines. Some of them remain pluripotent after prolonged cultures, while others became restricted in their differentiation potentialities. Sometimes they gave rise to a single committed cell type, which is the case for the so-called neuro-teratocarcinoma stem cells which differentiate in vitro only in neurones. Another ECC line, F9, differentiates into visceral endoderm if treated with retinoic acid (a vitamin A derivative) and into parietal extraembryonic endoderm if treated with a retinoic acid and cAMP mixture. Differentiation of F9 cells into visceral endoderm cells is 'accompanied by the activation of the a-fetoprotein gene. This gene remains silent in F9-derived parietal endoderm cells, which synthesize instead, plasminogen activator, cytokeratin filaments, laminin and type-IV procollagen.

Experimental Analysis of Cell Differentiation in Culture

Cell Fusion

A very interesting approach has been developed for the study of cell differentiation: it is the technique of *cell fusion,* also called *somatic hybridization*, which was first devised by Ephrussi. The properties of the cell membrane can be altered by treatment with a hemolytic virus, called the Sendai virus, even when this virus has been inactivated and is unable to multiply in the infected cells, or with polyethylene glycol. These changes in membrane properties allow the easy fusion of two (or more) cells of the same or different types. As Harris has done. when a nucleate hen red blood cell is fused with a human cancer cell (HeLa cell), very remarkable

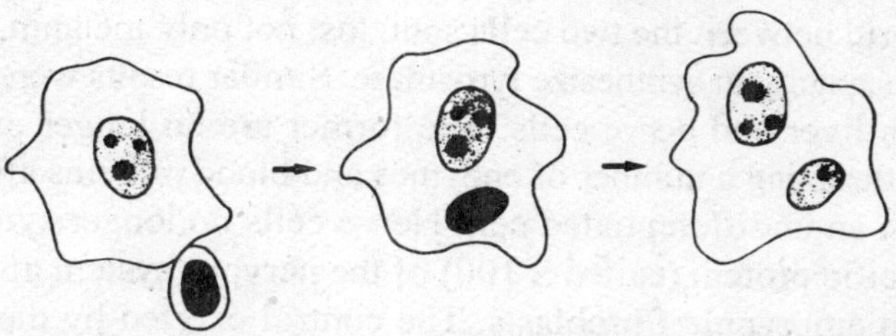

Figure 9.2 : Fusion of a chicken red blood cell (small, with a condensed nucleus) with a mammalian cell: the reactivated red blood cell nucleus forms a nucleolus.

changes occur. The nucleus of the red blood cell (erythrocyte) was, before fusion, in a completely inactive, repressed form. It synthesized neither RNA nor DNA; the HeLa cell nucleus, on the other hand, is extremely active in both DNA and RNA synthesis. When the two cells are fused together, a *heterokaryon is* produced. There are two nuclei, of different origins (hen and man) in a common cytoplasm. Analysis of this somatic hen-man hybrid shows that the erythrocyte nucleus is quickly *reactivated.* It becomes capable of synthesizing DNA and RNA. The hen erythrocyte nucleus contained no visible nucleoli before the cell fusion; after cell fusion, they become very conspicuous and RNA-rich. The nucleolar organizers are now active again. At that same time, it becomes possible to detect, by immunological methods, the formation of chickspecific antigens in the cell membrane of the somatic hybrid. The HeLa cell *cytoplasm* has thus exerted a positive. derepressing effect on the previously wholly inactive chick nucleus. This effect is entirely similar to the one we met before when the experiments of Gurdon on unfertilized *Xenopus* eggs were described. Injection of a nucleus taken from an adult organ (brain, liver), where there is no longer any DNA synthesis, into young cytoplasm is followed by a resumption of DNA synthesis.

Clearly, the same kinds of positive control that the egg cytoplasm exerts on the injected adult nuclei still operate in active adult cells.

This elegant technique of cell fusion has been used by Ephrussi in the study of cell differentiation. He fused together *melanocytes* and still undifferentiated embryonic cells *(fibroblasts)* and cultured these hybrids. The result, in this case, was that the undifferentiated cell exerted *a negative* control on the differentiated one. The hybrid between the two cells soon lost not only melanin, but even the capacity to synthesize tyrosinase. Similar results were obtained with liver and nerve cells. The former are no longer capable of synthesizing a number of enzymes and blood proteins after fusion with an undifferentiated cell. Nerve cells no longer synthesize a specific protein (called S 100) of the nervous system after fusion with embryonic fibroblasts. The control exerted by the undifferentiated cell on differentiated functions is not *always* negative but this is an exception to the general rule. Studies on the fusion *of* malignant and normal cells have shown that the situation prevailing in somatic hybrids can be very complex. Malignancy usually behaves like a recessive character, but the interpretation of the results is complicated by the fact that cell fusion followed by culture is often followed by the elimination of part of the chromosomes (this also often happens, as we have seen, in the ordinary sexual hybrids). Experiments, in Ephrussi's laboratory, have shown that if the melanocytes contain twice as many chromosomes as the fibroblasts, the ability to produce melanin is no longer lost in the hybrid between the two cells. This clearly shows the importance of "gene dosage" for the expression of the phenotype (melanin production).

Recent studies indicate that the rule, after cell fusion and culture, is the following: the differentiated phenotype (formation of pigment, liver enzymes, S 100 protein, etc.) is lost soon after cell fusion; but it reappears when a certain number of chromosomes have been lost during cell divisions in the hybrid line. The differentiated phenotype is first extinguished, then re-expressed. In order to gain some insight in the relative roles of the nucleus and the cytoplasm in the phenotypic expression of differentiation, cytoplasm from enucleated fibroblasts (called *cytoplasts)* has been fused with intact cells. In such *"Cybrids"*, extinction of the differentiated character does not last more than 12 to 20 h; the differentiated phenotype is restored 48 h after fusion. This experiment shows that a cytoplasmic factor exerts a negative control on cell differentiation, but that this factor has a short life and that it is not renewed in the absence of the nucleus. Since enucleation induces the differentiation of neuroblastoma cells (neurite formation), it is likely that the nucleus contains or produces factors which inhibit cell differentiation. This is another case of a negative control exerted by the nucleus on the cytoplasm. It is likely that extinction of the differentiated phenotype is due to changes in the molecular organization of chromatin. But it is too early to draw strong and general conclusions in a fast moving and complex field. We still know very little about the molecular interactions which take place between nucleus and cytoplasm in somatic hybrids and cybrids.

Effects of Bromodeoxyuridine (BrdUr) on Cell Differentiation

There is another interesting approach to the analysis of cell differentiation: Holtzer and colleagues found that an analogue of thymidine, bromodeoxyuridine (BrdUr) specifically inhibits cell differentiation in most cultured cells. At low concentrations, which do not stop DNA replication and cell division, BrdUr inhibits the in vitro differentiation of melanocytes, cartilage cells, muscle cells, etc. It has no remarkable effects on already differentiated cells and on cell viability, but it specifically inhibits the expression of the *luxury* proteins: melanin, chondroproteins, contractile proteins in respectively, melanoblasts, chondroblasts and myoblasts. It has also been reported that treatment of very young chick embryos with BrdUr suppresses the formation of red blood cells and the synthesis of hemoglobin. However, treatment with BrdUr of fertilized or cleaving sea urchin, ascidian, amphibian, hen and mouse eggs has led to disappointing results: all one can see is a slowing down of development, due to cell injury and death. There is no specific inhibition of tissue differentiation when the drug is added to whole eggs or embryos.

Why BrdUr exerts highly specific effects on the expression of the luxury proteins remains obscure at the molecular level.

Incorporation of the drug into replicating DNA obviously takes place, but it occurs randomly and we have only hypotheses for explaining the specific effects of the analogue on the genes which control the synthesis of the luxury protein. One possibility is that incorporation of BrdUr into DNA affects regulatory sequences controlling the activity of the structural genes coding for the luxury proteins. Another possibility is that certain chromosomal proteins bind specifically to the substituted DNA sequences: transcription of these sequences would then be shut off. Indirect effects on the cell membrane have also been suggested, but it is unlikely that they play a major role in the still mysterious BrdUr effect.

Effects of Phorbol Esters and Retinoids

The cancer problem lies outside the scope of the present book, but we cannot avoid mentioning it when cell differentiation is under discussion. Cancer is believed to be a multistep process (like embryogenesis!) leading to "mad" cells in which the regulatory mechanisms which control proliferation and differentiation in normal cells no longer function properly. All carcinogenic agents (physical, like X-rays; chemical, like the polycyclic hydrocarbons; viral, etc.) affect DNA. The cells in which DNA has been injured by a genotoxic agent will not necessarily become malignant: after *initiation* by a carcinogen, *promotion* by a co-carcinogen (promoter) is required as a second step. If. as shown by I. Berenblum almost 30 years ago, one applies locally a single dose of a carcinogen to the ear of a mouse, cancer will seldom develop, but if one rubs the ear of this mouse with croton oil, fast-growing tumors will form. Croton oil thus contains *a co-carcinogen* (promoter); the active principles of croton oil are *phorbol esters,* the more potent being 12-O-tetradecanoyl-phorbol-13-acetate (TPA).

At very low concentrations, TPA stimulates cell proliferation and, like BrdUr suppresses differentiation in cultures of myoblasts and chondroblasts. It is even able to induce the dedifferentiation of cultured myotubes and to disrupt selectively their myofibrils. Differentiation of cultured epithelial cells is also suppressed by TPA: they fail to synthesize their marker protein, keratin. In induced MEL cells, TPA inhibits hemoglobin synthesis.

So far, the effects of TPA on whole eggs and embryos have been seldom studied: the most striking, in both sea urchin and amphibian eggs is a dissociation of the superficial ectodermal cells. This suggests that the cell membrane is the initial site of action for TPA. There is now strong evidence for the view that TPA indeed acts on the *cell membrane:* in particular, cell-to-cell communication (i.e. the transfer of small molecular weight substances from a cell to its neighbors) is strongly decreased in TPA-treated cell cultures. Recently, it has been demonstrated that the cell membrane possesses specific receptors for phorbol esters. After binding to these receptors TPA activates a newly discovered membrane protein kinase (which requires phospholipids and Ca” for activity). It is now believed that this *protein kinase C* is the phorbol ester receptor. Its activation by TPA results in the phosphorylation of cytoplasmic proteins involved in the organization of the cytoskeleton and in protein synthesis.

It should be added that, as in the cases of cell hybridization and BrdUr treatment, the effects of TPA on cell differentiation are not always negative. For instance, TPA promotes the differentiation of a few human leukemic cell lines. We do not know why they make exceptions to the general rule; all one can say is that drugs like TPA and BrdUr exert “pleiotropic” effects, but this word masks our present ignorance.

It is interesting to note that the sites of action of BrdUr (DNA) and TPA (the cell membrane) are entirely different; yet the two drugs exert very similar effects on cell differentiation. Only more work will tell us whether TPA and BrdUr are ultimately acting on a single, still unknown, key step which would be absolutely required for cell differentiation. This step might perhaps turn out to be as simple as a change in the free calcium ion concentration or in the intracellular pH, since we know that such changes may have far-reaching consequences for the cell.

Derivatives of vitamin A (the *retinoids* related to the retina pigment retinene) often antagonize the promoting effects of phorbol esters on cancer cells. In certain malignant cell lines, *retinoic acid* (the most active of the retinoids) can induce the differentiation of undifferentiated cancer cells. These favorable effects on cell differentiation are linked to the fact that retinoic acid slows down the in vitro multiplication of malignant cells. The most striking effect of retinoic acid on cell differentiation has already been mentioned in this chapter: "nullipotent" (thus incapable of any kind of differentiation) embryonal carcinoma cells differentiate into endoderm if they are treated with retinoic acid. It is very likely that the action of the retinoids (which are liposoluble compounds) is mediated by a cytosolic retinoic acid binding protein (cRABP).

We do not know how the signals received by the cell, after treatment with either TPA or retinoic acid, are transmitted to chromatin where a few genes involved in cell differentiation must, in a selective way, be activated or shut down.

Other Inducers and Inhibitors of Cell Differentiation

It is likely that these signals are transmitted from the cell membrane to the nucleus, e.g. through *second messengers, cyclic* nucleotides or calcium ions. Increasing the cAMP content of the cells often slows down proliferation and favors differentiation. However, there are many exceptions to this rule. Similarly, there are numerous exceptions to the rule which states that a high cGMP content favors cell proliferation and prevents cell differentiation. As we have seen, a release of membrane-bound calcium ions triggers amphibian oocyte maturation and sea urchin egg fertilization. It is likely that changes in free Ca" and in internal pH play a role when cells stop dividing and begin differentiating. It is probable that cyclic nucleotides and calcium ions, when they trigger cell differentiation, activate protein kinases which can phosphorylate serine, threonine and tyrosine residues in proteins which are important for differentiation.

A few attempts to modify cell differentiation by the addition of so-called *conditioned media* to differentiating cells have been made. Conditioned media are culture media in which cells have been grown for a few hours or days. They contain substances which have been released or excreted in the surrounding medium by the growing cells. Conditioned media are valuable for the demonstration and isolation of substances which affect (positively or negatively) cell proliferation and of molecules involved in cell aggregation. However, no differentiation inducing substance has been isolated so far from conditioned media with the exception of a factor, present in the conditioned media of chondrocytes which stimulates the differentiation of these cells into cartilage. This factor, of unidentified chemical nature, stimulates the synthesis, by chondrocytes, of chondroproteins and collagen.

Popular today among the students of differentiation is *5-azacytidine,* which like BrdUr, is incorporated into replicating DNA molecules. The interesting property of this cytidine analogue is that it inhibits *DNA methylation. As* already mentioned, undermethylation of the cytidine residues in DNA molecules is a frequent - but not a constant - marker of "active" chromatin. In contrast to BrdUr, 5-azacytidine exerts *positive* effects on cell differentiation: for instance, it induces the differentiation of pre-adipocytes into adipocytes, of myoblasts into myotubes and of chondroblasts into cartilage cells. On the other hand, it has no effects (except sheer toxicity) on sea urchin and mouse eggs; unlike retinoic acid, it does not induce differentiation in nullipotent embryonal carcinoma cell lines. The most curious effect to date of 5-azacytidine is the *reactivation of the X chromosome* in females of the mammals.

Culture of a man-mouse hybrid cell line in the presence of 5-azacytidine is followed by reactivation of the X chromosome as shown by genetic analysis. However, only negative results have been obtained when normal diploid fibroblasts were treated with 5-azacytidine. Evidently, this substance is a promising tool for the analysis of cell differentiation; but more work is required before a close link between this process and DNA undermethylation can be accepted as a general fact.

EMBRYONIC DIFFERENTIATION AND CANCER

Although carcinogenesis and embryogenesis are two different topics, the two processes have so much in common that a short comparison is not out of order here. Embryonic differentiation is the last step prior to senescence and death; embryogenesis is a multistep process which starts at gametogenesis. Malignant transformation, as mentioned before, is also a multistep process.

It leads to the production of "transformed" cells, which can give rise to permanent cell lines. These cells are *immortal* and, in contrast to normal cells, never show signs of *senescence* (large size, low rate and finally arrest of DNA synthesis, modifications of the cytoskeleton, etc.). Instead, transformed cells go on dividing continuously. Although the initial lesion in the multistep carcinogenic process affects nuclear DNA, other cell constituents are also affected. In particular, the cell membrane properties are deeply modified in transformed cells. These cell surface changes allow transformed cells, in contrast to normal cells, to grow without attachment to a solid substratum (anchorage-independent growth). This, and the production by transformed cells of proteolytic enzymes, which can destroy the walls of the vessels, are responsible for the formation of the *metastases* which ultimately kill the host.

We have seen that our differentiated tissues derive from the proliferation of one or a few clones of cells; cancers also result from repeated division of a single malignant cell (clonal origin of cancers). All the cells which form a clone of normal, differentiated cells (for instance, all muscle or cartilage cells) look very much alike. This is not true for cancer cells: cells present in different metastases originating from the same tumor cell and even cells present in the same metastase display a striking *diversity. This* can be shown with immunological methods, which demonstrate a greater heterogeneity in malignant than in normal cells belonging to the same tissue. This heterogeneity might be due to genetic instability, leading to a high frequency of somatic mutations, in cancer cells. It is indeed well known that chromosomal abnormalities are frequent in cancer cells. However, the reasons why clones - in contrast to those of normal cells - of cancer cells undergo an increasing diversity when they expand by repeated cell divisions, remain mysterious.

It has often been suggested that cancer cells are more similar to embryonic cells than to adult cells. Cancer cells would result from the dedifferentiation of adult cells and this would allow them to proliferate repeatedly. This view is certainly an oversimplification of a more complex problem, but it cannot be disregarded. A classical example is the synthesis of two related proteins, a-fetoprotein and albumin, by liver cells. Embryonic liver synthesizes α-fetoprotein; around birth, the α-fetoprotein gene becomes silent; the albumin gene is switched on and albumin is synthesized during the whole lifetime. However, in liver tumors (hepatomas), the α-fetoprotein gene is reactivated and the albumin gene becomes inactive: thus, hepatomas synthesize an embryonic protein, α-fetoprotein. Another well-documented example is the frequent occurrence of *carcinoembryonic antigens* in tumors: these surface antigens are present in mammalian embryos and are no longer detectable in adults, but they are found in many tumors, in particular, in those of the colon. A particularly interesting case is that of the so-called F9 antigen of nullipotent malignant mouse embryocarcinoma (teratocarcinoma) cells. This antigen is shared with mouse sperm and morulae; it quickly disappears during development and it is absent from differentiated teratocarcinoma cells. The fact that teratocarcinomas may arise when early mouse embryos are grafted into an ectopic site and that malignant embryocarcinoma cells can undergo embryonic regulation, i.e. participate in normal development, after their introduction into a normal mouse blastocyst, show further that embryos and cancers must have much in common. The frequent reexpression of fetal genes in cancer cells has been called *retrodifferentiation*. However, there is an obvious difference between cancers and embryos: cancer cells proliferate and invade the host tissues in an anarchic way: embryos develop in a harmonious way because they are under the control of gradients, fields of differentiation and organizers. Positional information plays an essential role in normal morphogenesis; although it seems to be lost in neoplastic growth.

The origin of cancers has been the subject of extensive research in recent years. Since cancer is a multistep process *(initiation* by a genotoxic agent, *promotion* by agents which, like TPA induce a new phenotype, *proliferation* of the modified cells leading to a tumor by clonal expansion, formation and growth of *metastases),* it is impossible to pinpoint a single cause for all cancers. However, one of the factors which certainly plays an important role in malignant transformation, is the activation of cancer genes *(oncogenes),* which have been recently identified by molecular biologists.

It has been known, for many years, that a number of viruses induce tumors (leukemias, sarcomas) in birds and mammals (including man). Some of these viruses contain DNA, others RNA; the latter are called *retroviruses* because they contain an enzyme, *reverse transcriptase,* which copies the viral RNA into the complementary DNA which is integrated into the host genome as *a provirus.* Proviral DNA contains an oncogene responsible for malignant transformation. When these oncogenes are expressed in virus-infected cells, transforming factors are produced. In many cases, the viral oncogene product is a protein kinase, which phosphorylates several cellular proteins and alters their conformation, leading to the transformed phenotype. Recently, it has been found that certain viral oncogenes increase the production of growth factors or of their receptors on the cell surface. One of the major surprises in this fast-moving field has been the demonstration that *normal cells* contain oncogenes *(cellular oncogenes)* very similar to the viral oncogenes. Activation of normal cellular oncogenes could thus lead to malignant transformation. Today we know two possible mechanisms for *oncogene activation.* One of them is *point mutation:* replacement of a single base by another in a cellular oncogene will lead to the production of the homologous transforming viral oncogene. The products of the two oncogenes will differ by a single amino acid substitution and this might have farreaching consequences for the biological properties of the encoded proteins. The other mechanism for cellular oncogene activation is *translocation* of the gene from its normal site to a different site located on another chromosome. We have seen that chromosomal aberrations are frequent in tumors; among them are translocations in which a fragment of a chromosome is integrated in another chromosome. If a cellular oncogene is inserted in a chromosome region, which is favorable for transcription, it will be activated and the protein it is coding for will be expressed. Several examples in which cellular oncogenes are activated by translocation into the very active immunoglobulin locus (where antibodies are synthesized) are now known.

What is not known is the possible role of the cellular oncogenes in the normal cells which harbor them. These genes are silent in normal adult cells; but it is not excluded that they are active in the control of growth and differentiation in embryos. The fact that cellular oncogenes have been highly conserved during evolution and that they are present in *Drosophila* as well as in mice and men, indicate that they must have important functions. It will be an important task for the molecular embryologists of tomorrow to identify the products of the cellular oncogenes during development and to determine whether they play a role in proliferation and differentiation. If so, molecular embryology will benefit greatly from the recent progress made in cancer research.

THE FUTURE OF MOLECULAR EMBRYOLOGY

It would be absurd, for someone who has witnessed the stupendous progress made by molecular biology during the last 40 years to make predictions about the future development of molecular embryology. The discovery of a favorable material, which could be analyzed genetically, show good cell differentiation and lend itself to the preparation of in vitro systems, would be enough to make a big jump ahead. We need, among the eukaryotes, an organism which can be compared, in simplicity, with the bacteria and the viruses. Whether the ideal material for the study of cell differentiation really exists seems doubtful. All we can do now is to select systems that have advantages for the study of a particular problem and compare them. This is what we did, for example, when we compared *Acetabularia* and sea urchin eggs from the viewpoint of the morphogenetic and biochemical potentialities of nonnucleate cytoplasm. There is no doubt that the developing embryo, despite its complexity, remains the best material for the study of cell differentiation.

In the first edition of this book, we discussed a number of problems which seemed particularly important. We thought that their solution would be a turning point for molecular embryology, but this solution seemed so remote that they looked more like dreams than realities. Due to the astounding progress made by molecular biology, within little more than 10 years, these dreams have indeed become realities; molecular embryology has greatly profited - and will profit more and more in the future - from the availability of pure genes due to recombinant DNA technology.

Cloned genes (or the cDNAs of specific mRNAs) can be used for the detection, by in situ hybridization, of individual gene localization on metaphase chromosomes. We can now see, under the microscope, where specific genes (the hemoglobin genes, for instance) are located. Purified genes can be injected into the nucleus of a *Xenopus* oocyte where they will be accurately transcribed; these genes may be modified by chemical or enzymatic manipulation prior to their injection and this will tell us more and more about the control of transcription. Injection of mRNAs into the cytoplasm of *Xenopus* oocytes is now a standard method for the analysis of message translation. These oocytes are often a more convenient and efficient material than ribosomes isolated from bacteria, wheat germs or reticulocytes for those interested in the mechanisms of protein synthesis.

It is now possible, as we have seen, to inject pure genes into one of the pronuclei of a fertilized mouse egg and to obtain *transgenic* mice which express the injected gene. This seemed, 10 years ago, a very remote dream; the obtainment of transgenic mice is a major victory for molecular and experimental embryology. There is no doubt that we shall learn a good deal more about gene control in different tissues from this kind of experiment. It should be added, although the topic is outside the scope of this book, that the introduction of cloned, pure genes into somatic cells (fibroblasts, for instance) by addition of DNA to these cells is now a routine experiment: hundreds of genes have been introduced into cells by such *transfection* experiments and their fate (integration into the genome followed by replication, or degradation of the DNA molecules which have not been integrated) has been followed. In addition, somatic hybrids, in particular, those between mouse and human cells, have provided invaluable information about the localization of many genes on human and mouse chromosomes. Those who have witnessed the slowness of our progress in gene "mapping" on human chromosomes by classical genetic analysis of pedigrees are amazed to see that the mouse and human genomes might soon be known in as much detail as that of *Drosophila*.

We wondered, 12 years ago, whether the progress made in

mammalian embryology would lead to the production of "babies in test tubes". We know that they now exist; artificial insemination of human eggs is now routinely performed in many hospitals; culture of the fertilized eggs and implantation until birth in their own or a foster mother has been successfully achieved without ill effects so far. We wish to point out again the ethical problems raised by experimental work on human eggs and sperm: by properly selecting eggs and sperm, one could try to obtain a higher proportion of Nobel Prize winners or football players. By separating blastomeres during cleavage prior to implantation, one could obtain, at will, identical twins. When one is able to separate the spermatozoa which possess the X or Y chromosome, it will be possible to shift, at will, the proportion of men and women in the world. Injection of genes into a pronucleus of a fertilized human egg might lead to transgenic men, a rather frightening prospect. All this would be Huxley's *Brave New World,* a novel which would be a perfect anticipation of a future society, if the key word was not missing: DNA.

The day will come when, thanks to the development of more and more sophisticated machines for DNA sequencing, we shall know the whole base sequence of the human, mouse, *Drosophila,* sea urchins, etc. genomes. This will not explain embryonic development, but we shall know completely the information received by the fertilized egg, which will finally lead to the formation of an adult belonging to the same species. DNA is a unidimensional molecule and living things are three-dimensional beings: to understand morphogenesis, we must know much more about the tridimensional structure of proteins and other large molecules. From the DNA base sequence, the primary structure of the encoded proteins may be deduced, but not the interactions between the tridimensional structuresof these proteins with lipids or with other proteins. We doubt that, as was once humorously suggested by Sidney Brenner, if one fed all these data to a computer, a living mouse would come out of the machine (unless it was living in the computer).

Falling now from dreams to reality, we think that research in molecular embryology will go on very much as it does now: the experimental, reductionist approach, which has been so successful, will be followed for many years to come. Science progresses slowly: an experiment destroys a wrong idea, and this is progress; a new technology appears, often in a very different field, and this is another source of progress. All this leads to a better understanding of embryogenesis; perhaps the day will come when theoretical biology will be able to explain, in a mathematical formula, the unity and diversity paradox of the living world: but this is another dream and there are more urgent tasks ahead of us.

We should know more about gene regulation before we can understand embryonic differentiation; we should know much more about the molecular nature of the germinal localizations, of embryonic regulation and of inductions before we can indulge in theories. These are tasks for today and we do not doubt that outstanding progress will be made in the years to come. How long will it take before we understand how, at the molecular level, a hen egg becomes a chicken: a few decades, a few centuries? Perhaps we shall never know.

Index

D

E

U

V